Steel and Composite Structures

Steel and Composite Structures

Behaviour and design for fire safety

Y. C. Wang

CRC Press
Taylor & Francis Group
Boca Raton London New York

CRC Press is an imprint of the
Taylor & Francis Group, an **informa** business

A SPON PRESS BOOK

CRC Press
Taylor & Francis Group
6000 Broken Sound Parkway NW, Suite 300
Boca Raton, FL 33487-2742

First issued in paperback 2019

© 2002 by Taylor & Francis Group, LLC
CRC Press is an imprint of Taylor & Francis Group, an Informa business

No claim to original U.S. Government works

ISBN-13: 978-0-415-24436-7 (hbk)
ISBN-13: 978-0-367-86527-6 (pbk)

Typeset in Times by
Integra Software Services Pvt. Ltd, Pondicherry, India

British Library Cataloguing in Publication Data
A catalogue record for this book is available
from the British Library

Library of Congress Cataloging in Publication Data
Wang, Y. C. (Yong C.), 1964–
    Steel and composite structures: behaviour and design for fire safety/
      Y. C. Wang.
      p. cm.
    Includes bibliographical references and index.
      1. Building, Fireproof. 2. Building, Iron and Steel. I. Title.

TH1088.56 .W36 2002
693.8'2–dc21   2001054271

Visit the Taylor & Francis Web site at
http://www.taylorandfrancis.com

and the CRC Press Web site at
http://www.crcpress.com

To Mei Juan and our son Shi Long

Contents

Preface

Fire is one of the most fearsome natural phenomena and if not managed properly can lead to devastating consequences. Until very recently, measures to tackle fires in a building have followed procedures that have evolved over many years as reactions to previous disasters. These practises are enshrined in approved fire regulations and in various fire codes, collectively known as the prescriptive approach. It has been recognized by practitioners that strict adherence to these prescriptive rules can lead to uneconomic and inflexible design and from a desire to change things, the new "fire engineering" profession has been created. Steel structures are in a particularly advantageous position to benefit from advances in fire engineering. With a better scientific understanding of fire behaviour and its impact on steel structures, it is possible to devise methods to make the design and construction of steel structures more safe and economic. This vision has driven the steel fire research agenda that has led to major recent advances in this area. It is the author's belief that the steel industry is now standing at the threshold of much larger scale applications of fire engineering to benefit steel construction.

The author first became involved in fire engineering of steel structures in 1989 when he started his career as a research engineer at the Building Research Establishment to develop methods to analyse the behaviour of steel structures under fire conditions. This was the time when the first ever formal "fire engineering" code for steel structures, BS 5950 Part 8, was about to be published. His eight years at BRE was fruitful, particularly he had the opportunity to be part of the BRE team carrying out the large-scale steel structural fire research programme in the Cardington laboratory. The Cardington research programme is and probably will be the most important event in steel fire research and is already shaping the future of fire engineering of steel structures. In a few years since the Cardington test programme was completed, important new understandings have been obtained through the collaborative efforts of a number of organizations.

The idea of writing a book to summarize developments in this important area formed when the author started to teach fire engineering at the

University of Manchester in 1998. The preparation of lecture notes gave him the opportunity to conduct a systematic and critical assessment of a variety of sources of information in this area, particularly the post-Cardington advances. The particular emphasis of this book is to provide a detailed account of the fundamental behaviour of steel and composite structures in fire. It is only through an understanding of fundamental behaviour that new advancements in design can be made. This book is aimed at those who wish to have a better understanding of the behaviour of steel structures in fire and their influences on fire engineering design of steel structures.

Throughout his career, the author has had opportunities to interact with a large number of individuals and benefited enormously from discussions with them. In particular, this book draws much from the author's experiences gained at the Building Research Establishment and he is most sincerely grateful to help and assistance from many of his colleagues at BRE, especially, Dr David Moore, Mr Tom Lennon and Professor Haig Gulvanessian. The author also wants to thank Professor David Nethercot for introducing him to BRE at the start of his career and for continuously offering him support throughout his career so far.

It is inevitable that writing a book takes a lot of time and effort. This means sacrifice of other commitments. In this regard, the author wishes to thank his colleagues at the University of Manchester for allowing him to get away with doing so little and his research students for not having enough time to interact with them more often. In particular, he would like to express his gratitude to his colleague Mr M. R. Maidens for reading and making valuable comments on the manuscript. The people he is most indebted to are his family. His wife Mei Juan gave him motivation and encouragement to start this project in the first place and had the patience to support him throughout. During the writing of this book, their son Shi Long was born. He gives them great joy yet he was often not available to give him his full attention. Without their parents helping them look after Shi Long, it is doubtful whether he would have had the time or the will to complete this project. The author wishes to thank everybody very much.

Y. C. Wang
The University of Manchester
2001

Notations

a	Longer span of a slab
A	Area
A_C	Compression area
A_f	Floor area
A_{ff}	Area of fuel bed
A_g	Gross cross-sectional area
A_p	Perimeter surface area of steel section
A_s	Cross-sectional area of steel
A_T	Tension area, total surface area of fire enclosure
A_v	Area of opening
b	Shorter span of a slab, thermal property of lining material ($=\sqrt{k\rho C}$), width
B	Effective width of concrete flange
B_{eff}	Effective width of a thin-walled steel plate
C	Compressive stress resultant, specific heat
C_d	Orifice coefficient
d	Depth, distance between centroids
d_c	Depth of concrete in compression
D	Depth of enclosure
D_s	Depth of steel section
D_p	Depth of composite slab
E	Young's modulus
E_b	Blackbody radiant power
E_{cd}	Design modulus of concrete
EI	Stiffness
f_c	Compressive strength of concrete
f_{yr}	Design strength of reinforcement
g	Gravitational acceleration
G	Shear modulus
h	Heat transfer coefficient
h_c	Convective heat transfer coefficient
h_r	Radiant heat transfer coefficient

h_v	Opening height
h_w	Heat transfer coefficient at fire enclosure boundary
H	Height
I_n, I_θ	Intensity of thermal radiation
I_w	Warping constant
I_y	Second moment of area about the minor axis
J	Torsional constant
k	Thermal conductivity
$k_{E,T,com}$	Elasticity retention factor of steel at temperature of compressive flange
$k_{y,T}$	Effective strength retention factor of steel at temperature T
$k_{y,T,com}$	Effective strength retention factor of steel at temperature of compressive flange
K	Degree of shear connection
K_c	Stiffness of a restrained column
K_{co}	Initial stiffness of a restrained column
K_s	Restraint stiffness
L	Span of structural element
L_e	Buckling (effective) length
L_f	Total fire load
L_H	Horizontal length of flame
m	Modification factor for non-uniform bending moment distribution
m_{air}	Rate of air entrainment
m_{smoke}	Rate of smoke production
m_{wood}	Rate of burning of wood
M_b	Bending moment resistance for lateral torsional buckling
M_c	Bending moment capacity of a beam
M_{con}	Connection bending moment capacity
M_{cr}	Elastic lateral torsional buckling bending moment
M_{cx}	Plastic bending moment capacity about the major axis
M_{cy}	Plastic bending moment capacity about the minor axis
M_{fi}	Applied bending moment in fire
M_{max}	Maximum bending moment
M_p	Plastic bending moment capacity of a cross-section
M_{pc}	Plastic bending moment capacity of a composite cross-section with complete shear connection
M_s	Plastic bending moment capacity of a steel cross-section
M_x	Bending moment about the major axis
M_y	Bending moment about the minor axis
M_+	Sagging (positive) bending moment capacity
M_-	Hogging (negative) bending moment capacity
O	Opening factor $= \left(A_v \sqrt{h} / A_T \right)$
p_b	Bending strength of steel for lateral torsional buckling
p_c	Compressive strength of steel for flexural buckling

p_e	Elastic buckling stress
p_y	Design strength of steel at ambient temperature
P	Applied compressive load, perimeter length
P_c	Design compressive strength of a column
P_{cr}	Euler buckling load
P_{max}	Maximum axial load
P_0	Initial axial load
P_u	Squash load
Q_w	Latent heat of water
$q_{f,d}$	Design fire load density per unit floor area
$q_{t,d}$	Design fire load density per unit enclosure area
Q	Heat flux
r	Distance
r_y	radius of gyration
R	Load ratio, thermal resistance, burning rate of wood
R_r	Resistance of shear connectors
S	Plastic modulus of a cross-section
S_x	Plastic modulus of a cross-section about the major axis
t	Time, thickness
T	Temperature, tensile stress resultant
t_{eqv}	Equivalent time of fire exposure
T_w	Temperature of fire enclosure boundary
T_x	Catenary force
U_0	Velocity
V	Volume
w	Density of uniformly applied load
w_c	Water content
W	Width of enclosure
Z_y	Elastic modulus about the minor axis

Greek letters

α	Robertson constant for flexural buckling, coefficient of thermal expansion, absorptivity
χ	Strength reduction factor for buckling
χ_{LT}	Strength reduction factor for lateral torsional buckling
δ_{max}	Maximum deflection
δ_{th}	Thermal bowing deflection
ε	Emissivity
ε_{cr}	Creep strain
$\varepsilon_{c,T}$	Concrete strain at temperature T
$\varepsilon_{cu,T}$	Concrete strain at peak stress at temperature T
ε_m	Emissivity of material
ε_{mec}	Mechanical strain
ε_r	Resultant emissivity

ε_{th} Thermal strain
η_{LT} Perry coefficient for lateral torsional buckling
λ Slenderness
$\overline{\lambda}$ Relative slenderness (Eurocode definition)
λ_{LT} Lateral torsional buckling slenderness of a beam
$\overline{\lambda}_{\text{LT}}$ Relative lateral torsional buckling slenderness of a beam (Eurocode definition)
μ Absolute viscosity
ν Relative viscosity
ρ Density, reflectivity
σ Stefan–Boltzmann constant, stress
$\sigma_{\text{c,T}}$ Concrete stress at temperature T
τ Transmissivity
Φ Configuration factor

Subscripts
a   Ambient temperature
c   Concrete
fi   Fire
p   Fire protection
r   Reinforcement
s   Steel
T   Elevated temperature

Chapter 1

Introduction

1.1 BACKGROUND

Recent years have seen intensive activities and major advances in the field of steel structures in fire, including research activities to gain a thorough understanding of this subject and accelerating engineering applications of the research results. In this book, the term "steel structure" is used to refer to both steel and composite steel-concrete structures. Not so long ago, one would automatically accept that steelwork would need fire protection and use the fire protection thickness based on the standard fire resistance tests. This is following the prescriptive approach. The validity of this prescriptive approach is now being questioned and the availability of new understanding and knowledge is enabling safer and more economical design and construction of steel structures for fire safety within the framework of the "fire engineering" or "performance-based" approach. Of course, the concept of the "fire engineering" approach is not new, some 30 years ago, Pettersson *et al.* presented the first comprehensive treatment of fire engineering of steel structures (Pettersson *et al.* 1976).

The essential difference between the prescriptive method and the "performance-based" approach for a steel structure in fire can be described in a very simple way. In the prescriptive approach, one is trying to limit the temperature rise in steel to about 550 °C when exposed to the standard fire condition. This is based on the belief that at temperatures above 550 °C, a steel structure may not be safe and below 550 °C it is safe. In this assessment, the steel temperature is the end condition and this approach does not address the specific circumstances of the structure, which include the type of fire that the structure is likely going to experience, the consequences of fire exposure, the loading condition, the importance of the different structural elements and the interactions between them. In the "performance-based" approach, all these specific requirements are considered. The steel temperature is merely one of many design variables and the standard fire exposure one of numerous types of fire severity. Obviously, the fire engineering approach can be much more difficult to apply than the prescriptive one, however, the benefits of using this new method can be enormous.

Two factors are the main drivers of such a change of attitude: (1) the demand for more competitive steel construction from the steel industry; and (2) a desire to pursue a better understanding of the subject from the research community. Until recently, the cost of fire protection accounted for about 30% of the total cost of a steel structure (Robinson and Latham 1986). This represented a significant addition to the construction cost and put the steel structure at a disadvantage relative to other forms of construction, especially concrete construction whose performance in fire is often considered to be much superior. The steel industry identified the fire protection cost as one of the main obstacles limiting its market share and committed extensive resources to studying this subject with the aim to substantially reduce the cost of fire protection. On the other hand, the steel research community identified the effects of fire on the behaviour of steel structures as a relatively poorly researched area and started to devote much time and effort to this problem.

The activities of the steel industry and the steel research community reached the climax with the large-scale structural fire tests on an eight-storey steel-framed structure in the Cardington laboratory of the Building Research Establishment (BRE), United Kingdom in the mid-1990s (Martin and Moore 1997). Although due to inevitable financial restraints, only a few fire tests were carried out, what these few tests have revealed are major breakthroughs and will shape the future design of steel structures in fire. These tests have provided the momentum and some vital quantitative information to enable the transformation from accidental observations of superior performance of complete steel structures under fire conditions to practical applications. The results of the Cardington fire tests are being used as the basis of extensive research studies for developing the next generation of "performance-based" design methods for steel structures in fire.

Set against such a background, this book aims to achieve two objectives. The Cardington fire tests and subsequent research studies, together with the information prior to the Cardington research programme, form a large body of knowledge on the behaviour of steel structures in fire. However, this existing body of knowledge often appears as research papers scattered around in various academic journals and conference proceedings. The first objective of this book is to bring these different sources of information together in a single volume in a systematic manner so as to produce a reference point for future activities.

At the same time, the "performance-based" approach for steel structures in fire is being developed and adopted; "fire engineering" in general is gaining wider acceptance as a profession. The field of fire engineering is broad, encompassing a diverse range of activities, including structural engineering, building services engineering, mechanical and electrical engineering, economics, psychology and management. Any new profession will need a wealth of text books and references for its training. The second objective of this book is to contribute to the training of fire engineers by gaining a better understanding of the structural engineering part of this broad field.

Due to relatively new start of the fire engineering profession and the multi-disciplinary nature of this subject, until very recently, books on fire engineering were few and far between. Even at present, when many fire engineering books are available and many more are being planned, they tend to concentrate more on the "fire" side and their explanations of structural behaviour are often neither detailed nor up to date. The few available structural fire engineering books (e.g. Malhotra 1982; Lie 1992; SFPE 1995; Purkiss 1996; Buchanan 2001) are useful introductions to this topic, but they tend to emphasize only on design calculations of individual elements under the standard fire exposure and do not include many of the major recent developments on natural fire behaviour and whole structural performance. Moreover, these books cover different construction materials and their treatments of steel and composite structures are inevitably limited in depth. These books are best used as calculation tools to supplement or even replace the standard fire resistance testing of a steel or composite structural member. They are inadequate to provide more detailed guidance on broader issues of fire engineering of steel structures. The afore-mentioned report by Pettersson et al. (1976) is an excellent source of information but its contents need updating in many areas.

At this stage, it is important to note contributions of the Steel Construction Institute (SCI) in the United Kingdom in this area. The United Kingdom is rightly in the forefront of activities on steel structures in fire. As the authoritative voice of the UK's steel design profession, the role of the SCI is important in promoting and disseminating new developments in the field of steel structures in fire. They have published numerous books and technical reports on different aspects of steel structural design for fire safety (e.g. Lawson and Newman 1990, 1996). These books are usually the main sources of information for the steel design profession on matters of fire safety. However, their descriptions of the fundamental structural behaviour under fire conditions are in general qualitative only and not detailed.

The main emphasis of this book is to provide a comprehensive quantitative description of various aspects of the behaviour of complete steel structures in fire, how current design methods are reflecting these behavioural aspects and how they are likely to be improved in the future when new understandings are developed. It is not intended to provide examples of detailed design calculations, as such it should be seen as complementary to the existing "design" books.

1.2   LAYOUT

This book is divided into 10 chapters, organized around the two themes of behaviour and design of steel structures in fire.

Chapter 2 introduces the behaviour and design of steel and composite structural elements at ambient temperature. This chapter has been included because the ambient temperature behaviour and design methods are the basis of further discussions for the fire situation. For readers who are experienced in structural engineering, this chapter may be skipped.

In Chapter 3, the main observations from a range of different fire tests are described. Fire tests are carried out worldwide and they have been initiated to serve the different agendas of different investigators. Nevertheless, they appear to follow a reasonably clear common thread. This chapter starts from the relatively simple standard fire resistance test of a statically determinate individual element and continues to the large-scale structural fire tests at Cardington. These fire tests cover a range of structures with different complexity of structural behaviour in fire. The objectives of these descriptions are twofold:

1   to provide the reader with an introduction to the topics that have been analyzed by researchers and which will be included in this book; and
2   to reinforce the importance of considering the performance-based approach.

Chapter 4 provides a brief review of computer modelling of the behaviour of steel structures in fire. As with experimental studies, computer programs for the analysis of steel structures in fire have also grown in complexity. In the beginning, numerical procedures were developed to provide an alternative to tests so as to calculate the standard fire resistance time of a simple structural element. They now have the capability to simulate different types of structures, interactions between different structural members in complete structures and various advanced structural effects, such as membrane action and progressive collapse. They are also becoming more important in extrapolating the applicability of a limited number of fire tests and in providing information for the generation of new design guidance. Many computer programs have been developed by different researchers and a few of them are relatively well known within the steel fire research community. Since it is becoming increasingly possible for engineers to use these specialist programs as part of their design tools, it is useful to have an understanding of the assumptions and applicability of these computer programs. Chapter 4 provides a brief assessment of these programs. The result of this review can also serve to point out areas where future developments of these computer programs can be beneficial.

Experimental investigations are usually the first step of more detailed studies to develop new understandings, often aided by numerical tools such as those described in Chapter 4. Chapter 5 presents a quantitative description of these understandings. It starts with different modes of behaviour of individual steel members under flexural bending in fire, including yielding, local buckling, flexural and lateral torsional buckling, progressing to catenary action and membrane action at large deflections. A particular emphasis of this chapter is on interactions between different structural members in a complete structure. Wherever possible, design implications of these interactions are pointed out.

Chapters 3 to 5 can be loosely regarded as dealing with the "behavioural" aspects of steel structures in fire. The second part of the book is concerned with how to use this information in design. This book does not only provide a commentary on how to use the well-established design procedures in codes of practice such British Standard BS 5950 Part 8 (BSI 1990b) and Eurocodes 3 and 4 Part 1.2 (CEN 2000b, 2001), it also has the following additional objectives:

1   to compare the accuracy of different design methods;
2   to introduce simplified methods for easy implementation of some of the codified calculation procedures;
3   to assess the applicability of some of the codified design procedures to realistic fire conditions that have been derived on the basis of standard fire resistance tests;
4   to give guidance on how to include effects of structural interaction; and
5   to give a preview of some of the emerging new design methods following the Cardington experience which will take some time to be reflected in codes of practice.

The design chapters will deal with various aspects of steelwork design in fire. In general, the fire engineering design of a steel structure (more broadly, other structures as well) follows three steps:

1   The fire exposure condition is established. The result of this analysis gives the temperature–time relationship of the fire attack and defines the fire load on the structure.
2   Using the fire exposure condition as input, the temperature field in the structure is evaluated.
3   According to temperatures in the steel structure, the performance of the steel structure is then assessed. This assessment will determine whether the steel structure can meet its various fire resistant requirements.

All these three aspects will be addressed in this book. However, instead of starting with the fire behaviour, the first chapter on design (Chapter 6) will provide a brief introduction to heat transfer. Some knowledge of heat transfer analysis is necessary to understand the quantitative description of fire behaviour. Also since studies of heat transfer do not usually form part of the training programme of a structural engineer, this chapter is a useful introduction to familiarize structural engineers with the methodology and terms used in fire engineering calculations. Detailed treatment of heat transfer is complex and is beyond the scope of this book. This chapter will only describe basic concepts of heat transfer and introduce such heat transfer equations and assumptions as are relevant to heat transfer analysis of structures under fire conditions.

Chapter 7 deals with fire behaviour and the associated design assumptions. For more detailed treatments of fire dynamics, the reader may consult a number of excellent recent publications (e.g. Drysdale 1999; Karlsson and Quintiere 2000). This chapter will only provide a description of the aspects of fire behaviour that are relevant to structural design calculations. For the behaviour of a steel structure to be noticeably affected by fire, the severity of fire exposure (in terms of its temperature and duration) should be reasonably high. Therefore, this chapter will mainly describe the so-called post-flashover enclosure fires. In some cases, structural behaviour may also be affected when the structure is subjected to an intense localized fire. This book will also give some information on this aspect.

The design of steel structural elements in fire are presented in two chapters, Chapter 8 dealing with bare steel elements and Chapter 9 composite steel-concrete elements. For statically determinate structural elements, presentations in these two chapters mainly follow provisions in the established codes of practice, namely Eurocodes (CEN 2000b, 2001) and British Standards (BSI 1990b). These two sets of standard have been chosen to reflect the author's familiarity with them and also the fact that they are the two major codes of practice devoted to steel structures in fire. A structural element in a complete structure interacts with the adjacent structure. This interaction becomes complex when the structure is exposed to fire attack. Existing codes of practice do not deal with this behaviour in any detail. The design chapters of this book will describe how different structural elements in a complete structure may interact under fire conditions and how these interactions may be considered in fire resistant (FR) design.

As pointed out earlier in this introduction, the main reason for the steel industry's involvement in studying the performance of steel structures in fire is their commercial interest of using unprotected steelwork, thereby increasing the attractiveness of using steel and the steel market share. Therefore, it is not surprising that the main objective of their interest is in the elimination of fire protection to steelwork. The design methods in Chapters 8 and 9 may be developed to meet the above objective and Chapter 10 will provide a summary of how these may be done.

1.3 SCOPE

When planning for this book, the author has deliberately omitted some important topics in this area, including a detailed description of the standard fire resistance test, the specification and appropriate applications of fire protection materials, the assessment and repair of fire damaged structures and the economics of fire protection. The standard fire resistance test will continue to play an important role in the classification of construction elements for their fire performance. However, it is no more than a grading tool and there are many drawbacks in the set-up of standard fire resistance tests. In any case, as far as the structural behaviour and design are concerned, the standard fire exposure represents only one of numerous types of fire exposure that may be encountered by a structure. The selection of a particular fire protection system is relatively straightforward, once information is provided on the expected performance (e.g. in terms of limited temperature rise) of a structure that will use fire protection. In many cases, the role of a fire engineer or structural engineer may be to assess the condition of a fire damaged steel structure. The assessment of a fire damaged structure may be considered to be the reverse process of design for fire resistance and the information presented in this book will still be of some use. Moreover, the Corus group (formerly British Steel) have published a self-contained and comprehensive guide on the reinstatement of fire damaged steel structures and interested readers should consult this report (Kirby et al. 1986). A report published by the Institution of Structural Engineers (ISE 1996) on the appraisal of structures also provides a concise summary of this topic.

In deterministic design of structures under fire conditions, the worst-case scenario is adopted and the occurrence of fire and its development to the post-flashover phase are assumed to be a certainty. Therefore, adequate fire resistance is required to maintain safety of the structure. However, it should be realized that the occurrence of a fire is rare and is probabilistic in nature. Once ignited, whether or not a small fire can develop into the post-flashover stage to endanger structural safety depends on many factors, i.e. the supply of oxygen and combustible materials and early fire fighting. Moreover, the consequence of collapse of different buildings in fire can be different, some may be acceptable and the majority will not. Therefore, design for adequate fire resistance should ideally be based on risk assessment (CIB W14 1983); i.e. the level of the required fire performance of a structure should be related to the risk that is being presented by fire attack, taking into consideration the effectiveness and reliability of such fire safety provisions as fire detectors, sprinklers and fire brigade activity.

The risk-based approach has the potential to find the optimum trade-off between different active and passive fire protection methods. Although studies on the interaction of different fire protection methods have been going on for sometime, a systematic quantification is still not possible due to a lack of statistical data and the relatively short time devoted to this subject compared to that given to fire dynamics or structural performance. The work by Ramachandran (1998) represents perhaps the best effort in this direction so far and interested readers should consult his book.

This book is mainly intended for structural engineers who want to develop a good understanding of recent advances in the field of fire engineering related to steel structural design. Fire engineers will also find this book useful to help them develop an appreciation and some understanding of the complexity of the behaviour of steel structures in fire.

Since 1998, the author started to teach "Fire Engineering" to the final year students in Civil and Structural Engineering at the University of Manchester. The contents in this book form a large part of the lectures. This experience has benefited the author in many ways and has contributed to the selection of materials covered in this book. It is hoped that this book will be a useful addition to the references of other lecturers and students in this field and the author would be grateful to receive any constructive comment on how this book may be improved. This book attempts to summarize research activities in this area and it is also the hope of the author that this book can be used by his colleagues in the research community to help them formulate research ideas. Recent developments in this area are at such a pace that many things in this book will become out-of-date once printed and it is the author's intention to update this book in the future. In this regard, he would be most grateful to receive research publications from his colleagues in this field.

Chapter 2

An introduction to the behaviour and design at ambient temperature

As Chapters 3 and 5 will show the effects of fire on a steel structure are to make its structural behaviour more complicated than that at ambient temperature. To aid understanding of the fire effects and to provide a reference point for further discussions, this chapter provides a brief introduction to steel structural behaviour and relevant design methods at ambient temperature. For readers who are experienced in structural engineering and design, this chapter may be skipped. Therefore, this chapter is primarily intended for those readers who are less familiar with current UK and European codes of practice for the design of steel and composite structural members.

Current structural steel design codes of practice adopt the philosophy of limit state design. Each limit state is a condition beyond which a structure loses its intended function. The most obvious limit state is the ultimate load carrying capacity of the structure. If the load carrying capacity is exceeded, a structural collapse may occur. Other limit states include excessive deflections in beams or excessive lateral drifts in columns in normal service conditions, brittle fracture, fatigue, corrosion and durability. Of course, fire resistance is also a limit state. Of all limit states usually considered at ambient temperature, the ultimate limit state of load carrying capacity is the most relevant to fire resistant design and will be described in this chapter.

There are many uncertainties in structural design, among them uncertainties about loading conditions, material properties and calculation models. To accommodate these uncertainties, partial safety factors are introduced in design calculations. Exact values of partial safety factors are difficult to evaluate and the intention of using these partial safety factors is to ensure that the probability of structural failure is reduced to an acceptable limit. In design calculations, these safety factors are used to reduce the strength and stiffness of structural materials and to increase the applied load. Since these safety factors neither alter structural behaviour nor the basis of design calculations, they are omitted in this book to simplify presentation.

Beams, columns and connections are the main components of a building structure. Properly designed connections at ambient temperature are rarely a problem under fire conditions. Therefore, this chapter will only introduce the behaviour and design methods for calculating the ultimate strength of steel and composite beams and columns at ambient temperature. Presentations in this chapter will mainly follow the British Standard BS 5950 Part 1 (BSI 1990a), wherever possible, comparisons with Eurocodes will also be made.

2.1   LOCAL BUCKLING OF STEEL PLATES

Due to high strength to weight ratio, a steel cross-section is usually made of relatively thin plates. When a thin plate is under compression, local buckling may occur if the plate width to thickness ratio is too high. This is illustrated in Figure 2.1. Before local buckling, the steel plate is under uniform compression. After local buckling, stress distribution in the steel plate is no longer uniform and load is mainly resisted by areas of high stiffness. The design treatment for local buckling is to use an effective width for the steel plate, which is defined as the width of the steel plate that is subject to the maximum stress so as to have a stress resultant that is equal to the gross plate width, with the non-uniform stress distribution. The effective width concept is shown in Figure 2.2.

Local buckling is almost always encountered in thin-walled construction. In a hot-rolled steel section, steel plates are thicker and the problem of local buckling is less common, nevertheless, design calculations should always check whether local buckling is likely to be a problem. This is usually done by checking whether the ratio of the steel plate width to thickness exceeds a limit. If it does, local buckling is likely to occur, otherwise, local buckling does not need to be considered. The limiting width to thickness ratio of a steel plate without local buckling depends on the supporting conditions of the plate and Table 2.1 gives the limiting ratios for a few commonly encountered situations. For detailed information on how to calculate the

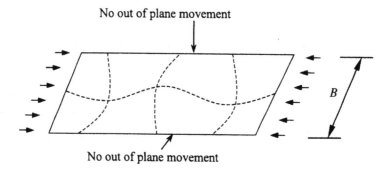

Figure 2.1 Local buckling in a thin-walled steel plate under compression with simply supported edges.

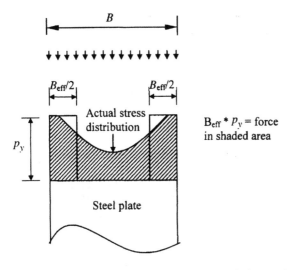

Figure 2.2 Effective width of a thin-walled steel plate under compression.

Table 2.1 Limits of steel plate width to thickness ratio for local buckling

Type of element	Width to thickness ratio
Flange of a rolled I/H section	15ε
Web of a rolled I/H section in pure compression	39ε
Web of a rolled I/H section in pure bending	120ε
Plate in a closed rectangular tube	39ε
Circular tube	$80\varepsilon^2$

Source: BSI (1990a). Reproduced with the permission of the British Standards Institution under licence number 2001SK/0298.

Note

$\varepsilon = \sqrt{275/p_y}$, where p_y is the design yield strength of steel.

effective width of a steel plate with local buckling, readers are referred to the British Standard BS 5950 Part 5 (BSI 1987b) or Eurocode 3 Part 1.3 (CEN 1996).

2.2   STEEL BEAMS

2.2.1   Plastic bending moment capacity

Under pure bending, a steel beam will most likely fail in one of two ways: (1) either by complete yield of the cross-section at the position of the

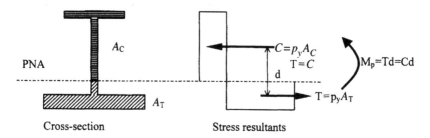

Figure 2.3 Determination of the plastic bending moment capacity of a steel cross-section.

maximum bending moment; or (2) by lateral torsional buckling (LTB) of the entire beam. If a beam is prevented from local and LTB, bending failure occurs when the maximum bending moment in the beam has reached the plastic bending moment capacity of the cross-section.

The plastic bending moment capacity of a steel cross-section is obtained from:

$$M_\mathrm{p} = p_y S \tag{2.1}$$

where p_y is the design strength of steel and S is the plastic modulus of the steel cross-section. For standard steel cross-sections, values of S can be found in steel manufacturers' section books. For a non-standard cross-section, the calculation procedure for M_p, as illustrated in Figure 2.3, is as follows:

1   The plastic neutral axis (PNA) is found. The PNA divides the cross-section into two equal areas, one in tension (A_T) and one in compression (A_C).
2   The plastic bending moment capacity of the cross-section is the resultant moment of these two equal forces, giving $M_\mathrm{p} = C \cdot d = \mathrm{T} \cdot d$, where d is the distance between the centroids of the tension and compression areas.

2.2.2   Lateral torsional buckling

The phenomenon of lateral torsional buckling of a steel beam occurs when the beam is loaded about its major axis, but deforms laterally accompanied by twist due to low torsional stiffness and low bending stiffness about the minor axis, see Figure 2.4. Lateral torsional buckling may be prevented by providing sufficient lateral and torsional restraints to the beam. Both types of restraint are effectively provided if the beam is laterally restrained at its compression side. This is usually the case in multi-storey composite

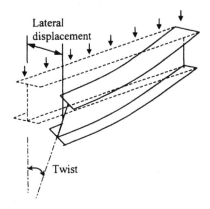

Figure 2.4 Lateral torsional buckling of an unrestrained beam.

construction when the compression (top) flange of a steel-floor beam is restrained by continuous floor slabs. Under this condition, it is not necessary to consider LTB. However, LTB should be considered when it is not practical to provide sufficient lateral and torsional restraints. Examples of inadequate lateral and torsional restraints include roof beams or near the support of a continuous beam.

When a beam fails in LTB, the maximum bending moment in the beam is much lower than the plastic bending moment capacity of the cross-section. For a simply supported, initially perfect, beam under a uniformly distributed bending moment, the elastic LTB resistance is given by:

$$M_{cr} = \frac{\pi}{L_e} \sqrt{EI_y GJ} \sqrt{1 + \frac{\pi^2 EI_w}{L_e^2 GJ}} \qquad (2.2)$$

where I_y is the second moment of area of the cross-section about the minor axis; GJ is the torsional rigidity; and I_w the warping constant of the cross-section. L_e is the unrestrained length of the beam.

The plastic bending moment capacity and the elastic LTB resistance are the two upper bounds on bending moment resistance of a steel beam. In realistic situations, due to yielding and initial imperfections in the beam (e.g. initial crookedness and residual stresses), the bending moment resistance of the steel beam is lower.

The effects of initial imperfections depend on a number of factors, the most influential being the slenderness of the beam. For a short beam, bending failure is governed by plastic yielding of steel and the beam's bending moment resistance is close to the plastic bending moment capacity of the cross-section. For a long beam, failure is by elastic buckling and the

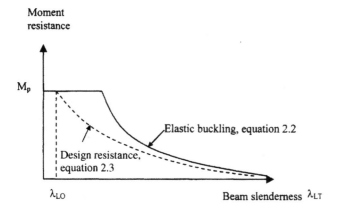

Figure 2.5 Bending moment resistance of a steel beam.

beam's bending moment resistance is close to the LTB resistance given by equation (2.2). For a beam with medium slenderness, failure is governed by elastic–plastic buckling and the beam bending moment resistance is much lower than either of the two upper bound solutions. Figure 2.5 sketches the variation of bending moment resistance of a steel beam as a function of its slenderness (λ_{LT}).

2.2.3 Design calculations according to the British Standard BS 5950 Part 1

2.2.3.1 Simply supported beams under uniform bending

According to the British Standard for steel structures BS 5950 Part 1 (BSI 1990a), the bending moment resistance of a steel beam under uniform bending is:

$$M_b = p_b S_x \tag{2.3}$$

where p_b is the bending strength of steel for LTB and S_x the plastic modulus of the cross-section about its major axis.

The bending strength p_b is related to the LTB slenderness λ_{LT} of the beam and the design strength of steel p_y. The LTB slenderness λ_{LT} is defined as:

$$\lambda_{LT} = \sqrt{\frac{\pi^2 E}{p_y}} \sqrt{\frac{M_p}{M_{cr}}} \tag{2.4}$$

where M_p and M_{cr} are defined in equations (2.1) and (2.2) respectively.

Calculating λ_{LT} using equation (2.4) is time consuming. In BS 5950 Part 1, λ_{LT} is calculated using:

$$\lambda_{LT} = uv\lambda \tag{2.5}$$

where λ is the flexural slenderness of the beam and is given by:

$$\lambda = L_e/r_y \tag{2.6}$$

in which L_e is the unrestrained length of the beam and r_y the radius of gyration of the cross-section about the minor axis. u is the buckling parameter and is a property of the cross-section. For a rolled I or H section, u is approximately 0.9. v depends on the slenderness ratio of the beam. For a symmetrical I or H section, the analytical expression for v is:

$$v = \left(1 + \frac{1}{20}\left(\frac{\lambda}{\chi}\right)^2\right)^{-\frac{1}{4}} \tag{2.7}$$

where χ is the torsional index of the cross-section and can be found in steel manufacturers' section book.

Having obtained λ_{LT}, the value of p_b can be obtained from a table in BS 5950 Part 1 according to the design strength of steel p_y. The relationship between p_b and λ_{LT} is often referred to as the beam buckling curve.

Alternatively, equation (2.4) can be used to obtain the elastic LTB stress as:

$$p_e = \frac{\pi^2 E}{\lambda_{LT}^2} \tag{2.8}$$

The bending strength p_b for LTB can be found from:

$$p_b = \frac{p_e p_y}{\phi_B + \left(\phi_B^2 - p_e p_y\right)^{1/2}}, \quad \text{with} \quad \phi_B = \frac{p_y + (\eta_{LT} + 1)p_e}{2} \tag{2.9}$$

where η_{LT} is the Perry-Robertson coefficient for lateral torsional buckling, accounting for the effects of imperfections on LTB resistance. For rolled sections, η_{LT} is obtained from:

$$\eta_{LT} = \alpha_b(\lambda_{LT} - \lambda_{LO}) \quad \text{but} \quad \eta_{LT} \geq 0 \tag{2.10}$$

in which $\alpha_b = 0.007$ is a constant, expressing the severity of initial imperfections. $\lambda_{LO}(=0.4\pi\sqrt{E/p_y})$ is the LTB slenderness of the steel beam below which LTB does not occur (see Figure 2.5).

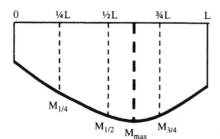

Figure 2.6 Bending moment distribution in a beam.

2.2.3.2 General design equations

If a beam is not loaded in uniform bending, stresses in the beam will not be at the maximum value everywhere. Therefore, the beam is less prone to LTB than under uniform bending. To take advantage of this, the maximum bending moment in the beam may be reduced so that the design check becomes:

$$\overline{M} = mM_{max} \leq M_b, \quad m \leq 1 \tag{2.11}$$

where m is a modification factor. Its value depends on the bending moment distribution in the beam and is given by (Taylor 2001):

$$m = 0.2 + \frac{0.15M_{1/4} + 0.5M_{1/2} + 0.15M_{3/4}}{M_{max}} \tag{2.12}$$

where M_{max}, $M_{1/4}$, $M_{1/2}$ and $M_{3/4}$ are the signed maximum bending moment and bending moments at $^1/_4$, $^1/_2$, and $^3/_4$ positions of the beam, see Figure 2.6.

Additionally, the maximum bending moment in the beam should not exceed the plastic bending moment capacity of the cross section, i.e.

$$M_{max} \leq M_p \tag{2.13}$$

2.2.3.3 A comparison between British Standard and Eurocode

Design calculations for the bending moment resistance of a steel beam in the European code Eurocode 3 Part 1.1 (CEN 1992a) are similar to those in the British Standard BS 5950 Part 1 (BSI 1990a). The only difference is in the beam buckling curve. Figure 2.7 compares the beam buckling curves obtained from these two design methods. It can be seen that for a beam

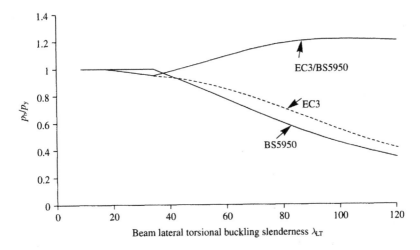

Figure 2.7 Comparison of beam lateral torsional buckling resistance from BS 5950 Part I (BSI 1990a) and Eurocode 3 Part 1.1 (CEN 1992a). Reproduced with the permission of the British Standards Institution under licence number 2001SK/0298.

where LTB governs, Eurocode 3 Part 1.1 predicts a higher buckling resistance, typically about 20% higher than that from BS 5950 Part 1.

2.3 STEEL COLUMNS

Columns are mainly subjected to compression loads. Under pure compression and in the absence of local buckling, a steel column may fail in one of two ways: (1) by complete yield of its cross-section; or (2) through loss of stability due to flexural buckling. The column slenderness is the most important factor to determine the failure mode of a column. A short column of low slenderness fails by complete yielding of steel in compression and the failure load is given by:

$$P_u = p_y A_g \tag{2.14}$$

where A_g is the gross cross-sectional area of the column. P_u, the compressive resistance of the cross-section, is often referred to as the column squash load.

For a long column, flexural instability governs and the upper bound capacity is given by the well-known Euler buckling load:

$$P_{cr} = \frac{\pi^2 EI}{L_e^2} \tag{2.15}$$

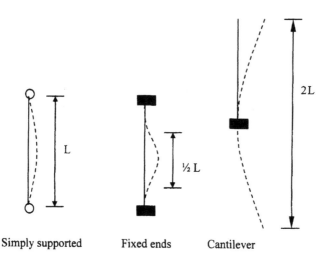

Figure 2.8 Effective length of a column with ideal supports.

From equation (2.15), the elastic buckling stress can be obtained as:

$$p_e = \frac{\pi^2 E}{\lambda^2} \tag{2.16}$$

where E is the Young's modulus of steel and I the second moment of area of the steel cross-section about the relevant axis of buckling. L_e is the column buckling or effective length, depending on the column support conditions. Figure 2.8 gives the effective length of a column for three common cases of ideal end supports: simply supported at both ends; encased at both ends and cantilevered at one end. The effective length of a column in a continuous frame depends on the relative stiffness of the adjacent structure to that of the column at both ends. Interested reader should consult BS 5950 Part 1, which is based on the work of Wood (1974). The slenderness, λ, of the column is given by equation (2.6).

Similar to LTB of a steel beam, flexural instability of a steel column is affected by initial imperfections. Therefore, the realistic buckling load of a column is less than those given by equations (2.14) and (2.15). The design resistance of a steel column is given by:

$$P_c = p_c A_g \tag{2.17}$$

Where p_c is the steel compressive strength. The value of p_c depends on the column slenderness and the relationship between p_c and λ is referred to as the column buckling curve. The analytical equation of a column buckling curve is:

$$p_c = \frac{p_e p_y}{\phi + \sqrt{\phi^2 - p_E p_y}} \quad \text{with} \quad \phi = \frac{p_y + (\eta + 1)p_e}{2} \quad \text{and} \quad \eta = 0.001\alpha(\lambda - \lambda_0)$$

(2.18)

where λ_0 $\left(=0.2\sqrt{\pi^2 E/p_y}\right)$ is the limiting slenderness below which column failure is governed by steel yielding.

The column buckling curve is affected by the effects of initial imperfections and the parameter η in equation (2.18) is used to account for this influence.

Different steel sections have different levels of imperfection, e.g. residual stresses in a universal column section are higher than those in a tubular section. Also, the same imperfection can have different effects on a column depending on the axis of buckling. Consider a universal cross-section as shown in Figure 2.9. During the rolling process, compressive residual stresses are generated at flange tips and tensile stresses at the flange/web junctions. Under the externally applied compressive load, compressive residual stresses can make the flange tips yield and lose their effectiveness earlier than the flange/web junctions. Whilst this reduces the column stiffness about the major axis linearly, reduction in the column stiffness about the minor axis is related to a cubic function of the distance from the flange tip to the web. Hence, the relative reduction in the column stiffness about the minor axis is much higher and the effect of residual stresses is much more severe about the minor axis than about the major axis.

The Robertson constant α is used to allow for the above mentioned different effects of initial imperfections on the column buckling load. Four values are given in BS 5950 Part 1, resulting in four column buckling curves, "a", "b", "c" and "d". The values of α are given in Table 2.2 for these four column buckling curves. Figure 2.10 compares the four column buckling curves with the steel yield strength and the Euler buckling stress of a perfect column.

A comparison between the British Standard BS 5950 Part 1 and the European Standard Eurocode 3 Part 1.1 will indicate that the two design methods give almost identical results.

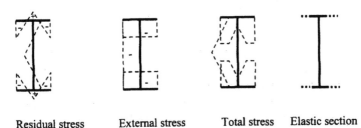

Residual stress   External stress   Total stress   Elastic section

Figure 2.9 Effect of residual stresses on the stiffness of a compression member.

Table 2.2 Robertson constant for column buckling curves

Column buckling curve	"a"	"b"	"c"	"d"
α	2.0	3.5	5.5	8.0

Source: BSI (1990a). Reproduced with the permission of the
British Standards Institution under licence number 2001SK/0298.

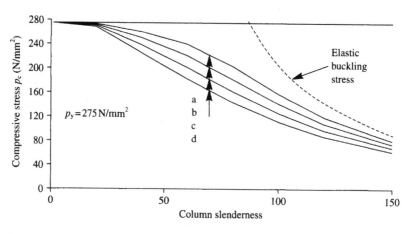

Figure 2.10 Column buckling curves (from BSI 1990a). Reproduced with the permission
of the British Standards Institution under licence number 2001SK/0298.

Comparing equation (2.9) for LTB of a beam and equation (2.18) for
flexural buckling of a column, it is clear that both types of buckling are
treated in a very similar way. This is not surprising since LTB of a beam is
in fact induced by flexural buckling of the beam plates that are under
compression. This is also why it is much more effective to restrain the
compression flange of a beam in order to prevent LTB.

2.4   COMBINED AXIAL LOAD AND BENDING

In addition to compressive load, a column may also be subject to bending
moments. In simple construction, the column bending moments come from
eccentricity. In continuous construction, the column bending moments are
transferred from the adjacent structure.

In BS 5950 Part 1, design calculations are carried out so that the local
capacity of the cross-section is not exceeded anywhere and that the column
does not lose its global stability. To carry out these design checks, axial
load – bending moment interaction equations are used.

The interaction equation for local capacity check is:

$$\frac{P}{P_u} + \frac{M_x}{M_{cx}} + \frac{M_y}{M_{cy}} \leq 1 \tag{2.19}$$

The interaction equation for global buckling capacity check is:

$$\frac{P}{P_c} + \frac{m_x M_x}{M_b} + \frac{m_y M_y}{p_y Z_y} \leq 1 \tag{2.20}$$

In these equations, M_x and M_y are the applied bending moments about the major and minor axes of the column at the critical cross-section. M_{cx} and M_{cy} are the plastic bending moment capacities of the cross-section about the major and minor axes. P_u, P_c and M_b are obtained from Equations 2.14, 2.17 and 2.3 respectively. Z_y is the elastic modulus of the cross-section about the minor axis. m_x and m_y are used to account for the effect of non-uniform distribution of bending moments in the column. For a column with linear distribution of bending moments, m is obtained from:

$$m = 0.57 + 0.33\beta + 0.10\beta^2 \quad \text{but} \quad \beta \geq 0.43 \tag{2.21}$$

where β is the ratio of the numerically smaller end moment to the larger one.

2.5 COMPOSITE BEAMS

A composite beam is formed by combining steel and concrete together. The most common type of composite beam is a steel beam connected to concrete slabs on top using shear connectors, shown in Figure 2.11. The concrete slabs act as the compression flange of the composite beam.

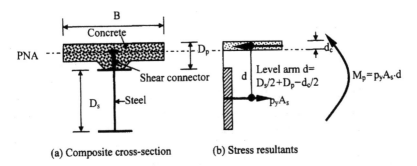

(a) Composite cross-section   (b) Stress resultants

Figure 2.11 Determination of the plastic bending moment capacity of a composite cross-section.

For a simply supported composite beam under gravity loading, the compression flange of the steel beam is laterally and torsionally restrained by the concrete slabs. Therefore, LTB of the steel beam is prevented. When designing the composite beam, it is only necessary to ensure that the maximum applied bending moment (M_{max}) does not exceed the plastic bending moment capacity of the composite cross-section M_p.

As shown in Figure 2.11, the plastic bending moment capacity of a composite cross-section is calculated in the following way:

1   The PNA of the composite cross-section is found. The PNA divides the composite cross-section into a tension part below and a compression part above the PNA. For pure bending, the tension and compression forces are equal. If the PNA is in concrete, it is assumed that the concrete in the tension zone below the PNA does not have any tensile strength. It is also assumed that the concrete in the compression zone is at a uniform stress that is equal to its design strength.
2   The plastic bending moment capacity of the composite cross-section is obtained by taking moment of the tension and compression forces in the composite cross-section.

For a composite cross-section with realistic dimensions, the PNA is usually in concrete so that the depth of the concrete in compression is:

$$d_c = \frac{p_y A_s}{f_c B} \tag{2.22}$$

where f_c is the design compressive strength of the concrete for bending and B is the effective width of the concrete flange. As an approximation, the effective width of the concrete flange may be taken as $L/4$ for an interior beam and $L/8$ for an exterior beam, where L is the span of the beam.

Using equation 2.22, the plastic bending moment capacity of the composite cross-section can be obtained using:

$$M_{pc} = p_y A_s \left(\frac{D_s}{2} + D_p - \frac{d_c}{2} \right) \tag{2.23}$$

2.5.1   Partial shear connection

In previous calculations, it is assumed that there is complete shear connection between the steel and concrete in a composite beam, i.e. there is no slip between the steel beam and the concrete flange at their interface. Complete shear connection is achieved by using a sufficient number of shear connectors so that the full force is transferred from the steel beam to the concrete flange or vice versa. Sometimes it may not be possible or necessary to achieve complete shear connection. In these cases, design for partial shear connection is possible.

For the case of PNA in concrete, complete shear connection is achieved when the total shear connector resistance is not less than the tension capacity of the steel section. The total shear connector resistance is measured from the point of zero bending moment to the point of the maximum bending moment in the beam. For a simply supported beam under symmetrical loading, the total shear connector resistance is calculated from half of the beam span. When the number of shear connectors is not sufficient such that the total shear connector resistance is less than the tension capacity of the steel beam, design calculations have to assume partial shear connection. Assuming that the total shear connector resistance is R_r, the degree of shear connection is defined as:

$$K = \frac{R_r}{p_y A_s} \le 1.0 \tag{2.24}$$

Under partial shear connection, the plastic bending moment resistance of a composite cross-section is less than that given by equation (2.23). Although it is possible to derive accurate analytical equations for the plastic bending moment capacity of a composite cross-section with partial shear connection, in most design calculations, the following simple and conservative inter-polation equation may be used:

$$M_p = M_s + K(M_{pc} - M_s) \tag{2.25}$$

where M_s is the plastic bending moment capacity of the steel cross-section (equation (2.1)) and M_{pc} is obtained from equation (2.23). Figure 2.12 indicates the difference between equation (2.25) and the exact method.

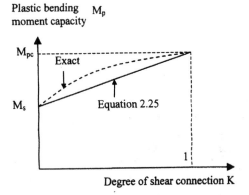

Figure 2.12 Influence of partial shear connection on the plastic bending moment resistance of a composite cross-section.

2.6 COMPOSITE COLUMNS

As shown in Figure 2.13, composite columns can be made in many ways. The earliest type of composite column is made by encasing a steel section in concrete to provide fire protection to the steelwork. Recently, concrete filled hollow steel sections are becoming more popular due to their load bearing efficiency, attractive appearance, speed of construction and inherently high fire resistance. Another type of composite column is obtained by filling concrete within the flanges of the steel cross-section. No temporary form-work for concrete casting is necessary. To make this type of composite column, the steel column is placed on its two flange tips and the concrete is cast on one side. After consolidation, the steel column is turned over and the concrete is cast on the other side of the column.

Design guidance for composite columns is provided in Eurocode 4 Part 1.1 (CEN 1992a) and BS 5400 Part 5 (BSI 1985). In both codes of practice, the calculation procedure for composite columns under combined axial compression and bending moments is complex. However, as will be described in more detail in Chapter 5, bending moments in a column in fire only play a secondary role. The following section will only describe design calculations for a composite column under pure compression. For detailed treatment of a composite column under combined axial load and bending moments, interested readers should consult the full design codes.

In fact, design calculations for a composite column under pure compression are identical to those for a steel column, the only difference being that in the case of the composite column, the squash load and rigidity of the composite cross-section should be used.

The squash load of a composite cross-section is given by:

$$P_u = A_s p_y + 0.85 A_c f_c + A_r f_{yr} \tag{2.26}$$

where A_s, A_c and A_r are the areas of steel, concrete and reinforcement and p_y, f_c and f_{yr} are their design strengths respectively.

Encased steel　　Concrete filled HSS　Partial encasement

Figure 2.13 Common types of composite column cross-section.

The constant 0.85 is used partially to account for the long-term effect of concrete exposed to environment. For concrete filled sections, this constant may be ignored so that:

$$P_u = A_s p_y + A_c f_c + A_r f_{yr} \tag{2.27}$$

In a concrete filled column, the concrete may be under confinement due to the steel tube restraining the lateral dilation of the failing concrete. This may increase the squash load of the composite cross-section. However, this benefit can only be realized when the composite column is short. In most realistic cases, this enhancement is small and may be safely ignored. In addition, under fire conditions, the steel tube is at a higher temperature and expands faster than the concrete. Any confinement effect is further eroded.

The rigidity of a composite cross-section is obtained from:

$$(EI)_e = E_s I_s + E_{cd} I_c + E_r I_r \tag{2.28}$$

where E_s, E_{cd} and E_r are the design modulus of elasticity of steel, concrete and reinforcement respectively. I_s, I_c and I_r are the second moment of area of these three components.

The same column buckling curves used for steel columns are also used in the design of composite columns.

2.7 PLASTIC DESIGN OF CONTINUOUS BEAMS

If a beam is continuous over a number of spans and its LTB is prevented, complete yielding of one cross-section of the beam does not indicate imminent failure. For example, for the continuous beam shown in Figure 2.14, the support bending moment will be higher than that in the span. When the support has completely yielded, the continuous beam does not fail. The effect of complete yielding is to form a plastic hinge in the support. Provided

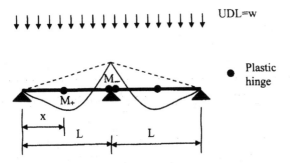

Figure 2.14 Failure mechanism of a two-span continuous beam.

the support can retain its plastic bending moment capacity and keep rotating, further loading in the structure is possible. Bending failure will occur only when more plastic hinges are formed in the beam span to develop a mechanism.

Under a uniformly distributed load, the distance of a span plastic hinge to the edge support is obtained from:

$$x = \frac{L}{2} - \frac{M_-}{Lw} \tag{2.29}$$

where M_- is the plastic bending moment capacity of the support cross-section (hogging bending moment capacity).

By static equilibrium, equation (2.30) is obtained to give the maximum load when a plastic hinge mechanism is formed:

$$M_+ + \frac{1}{2}M_- = \frac{1}{8}wL^2 + \frac{M_-^2}{2L^2w} \tag{2.30}$$

where M_+ is the plastic bending moment capacity of the span cross-section (sagging bending moment capacity).

Approximately:

$$\frac{1}{8}wL^2 = 0.45M_- + M_+ \tag{2.31}$$

The load carrying capacity of a simply supported beam under uniform loading is calculated from:

$$\frac{1}{8}wL^2 = M_+ \tag{2.32}$$

It is clear that due to contribution of the support, a continuous beam can resist much higher loads than a simply supported beam.

2.8 SEMI-RIGID DESIGN APPROACH

In conventional design, the connection between a steel beam and a steel column is assumed to be either a pin where no bending moment is transmitted or rigid where no relative rotation is allowed between the beam and the column. The performance of realistic connections lies somewhere in between these two extremes and is termed "semi-rigid".

Results (Nethercot 1985) of numerous research studies indicate that even the most flexible realistic connection has some capacity to transmit bending moment from the connected beam to the connected column. Since

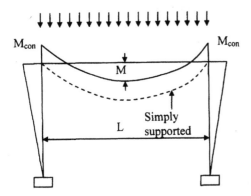

Figure 2.15 Effect of semi-rigid connections on beam bending moments.

these connections are usually idealized as pins in conventional design, it is possible to reduce structural cost by taking into consideration their actual contributions.

Figure 2.15 shows the effect of considering connection bending moment resistance for a beam. Since bending moments are resisted by realistic connections at both ends, the mid-span bending moment of the beam is reduced compared to a beam with pinned beam to column connections. Thus, under a uniformly distributed load, the maximum bending moment in the beam span is:

$$M = \frac{1}{8}wL^2 - M_{con} \qquad (2.33)$$

where M_{con} is the bending moment transferred by the connection.

Of course, the connection bending moment should also be considered in the design of the connected column. Compared to a column with a pinned beam–column connection, this additional connection bending moment will increase the column load. However, the bending stiffness of the realistic connection will also help reduce the effective length of the column, thereby increasing its resistance. In a comprehensive numerical study by Gibbons *et al.* (1993), it was shown that for realistic connections, it is always safe to design a column as simply supported without including the bending moment transmitted by the connection.

Chapter 3

Experimental observations

Understanding any aspect of the behaviour of a structure always starts from observations of the behaviour of physical models, either from carefully planned and executed experiments or uncontrolled accidents. This is also true of the behaviour of steel structures in fire. Whilst observations from accidental fires in steel structures can lead to some qualitative speculations of the possible behaviour of steel structures under fire attack, more precise and quantitative understanding can only be obtained from carefully planned and conducted experimental studies.

To date, many fire tests on steel structures have been carried out and this chapter presents a review of these fire tests. Such a review is necessary for a number of reasons:

- it will help to enumerate the modes of behaviour that exist in steel structures in fire, which will give directions to more detailed theoretical studies;
- even though computer simulations of steel structures under fire conditions are becoming more widely used, it is important that these numerical tools have the capability and accuracy to deal with all the likely modes of behaviour of realistic steel structures in fires; and this can only be done by checking theoretical predictions against fire test results; and
- a study of previous fire tests enables an identification of gaps of knowledge so that future experimental studies may be carried out more effectively.

Presentations in this chapter will start from the relatively simple behaviour of a simply supported steel column, through other types of structural elements to fire tests on complete structures. Of course, the behaviour of a structure in fire is affected by the type of fire exposure and the heat transfer process. However, due to the fact that steel and composite structures do not have strong interactions with the fire or heat transfer process, the aforementioned three areas are usually treated independently. The main focus of this review is on the structural (mechanical) behaviour. The related fire and thermal behaviour will be mentioned only if they become relevant to understanding the structural behaviour.

Numerous fire tests have been carried out on different types of structure. It will be impossible to provide a complete survey of all fire tests in this chapter. The fire tests selected for review in this chapter are restricted to those that reveal some special features of the behaviour of steel structures under fire conditions.

3.1   GENERAL TEST PROCEDURE

The testing of a loaded structure under fire conditions can be carried out in two generic ways: (1) transient state testing; or (2) steady state testing. In transient state testing, loads are applied to the structure first. These loads are then held constant and the structure is exposed to fire attack. The test is terminated when one of the specified failure criteria is reached. In steady state testing, the temperature in the structure is raised to the pre-determined level and held constant. Loads are then applied to the structure until structural failure. This is similar to structural testing at ambient temperature. If the structural behaviour is independent of the heating rate or the loading history, both methods of testing should yield the same result. However, this is not usually the case. Since fire tests are carried out using either one or the other method but not together, it is difficult to assess the difference between the two test methods. In this review, unless it is explicitly stated that the fire test was carried out under the steady-state condition, transient state testing should be assumed.

3.2   STANDARD FIRE RESISTANCE TESTS

Although structural fire tests must have been carried out before the invention of the standard fire resistance test, in the context of current knowledge, the most basic form of fire tests is the standard fire resistance test of a statically determinate simply supported structural element. The standard fire resistance test is usually carried out to assign a fire resistance rating to a construction element to enable it to pass the regulatory requirements for fire resistance. It is a device to grade the relative fire performance of different structural elements. So far, numerous standard fire resistance tests of steel structural elements have been carried out and some will be described in this chapter. Since the standard fire resistance test has been used as the basis of assessing the fire resistance of construction elements and a large body of knowledge have already been gained on the behaviour of steel structures under this type of fire exposure, it is useful to give a brief introduction to this test methodology so as to critically assess its relevance to understanding the behaviour of steel structures under more realistic fire conditions.

3.2.1   Test methodology

The standard fire resistance test is carried out according to a specified standard. In the United Kingdom, this is the British Standard BS 476, Part 20 (BSI 1987a). Other countries have their own standards. However, all these standards are similar. The international standard is ISO 834 (ISO 1975). For this reason, the standard fire exposure is often referred to as the ISO fire.

The standard fire test is carried out in a furnace, either gas or oil fired. Depending on the type, number, size and locations of burners in the furnace, different degrees of non-uniformity of temperature distribution will exist in the furnace. However, it is assumed that the combustion gas temperature inside the furnace is uniform and equal to the average temperature recorded by a number of control thermocouples inside the furnace. The average temperature rise is according to the following temperature–time relationship:

$$T_{fi} = 345 \log(8t + 1) + T_a \tag{3.1}$$

where the fire and ambient temperatures T_{fi} and T_a are in °C and the fire exposure time t is in minutes.

The standard fire test furnace is either a horizontal one, suitable for testing beams and slabs or a vertical one, used for testing columns and wall panels. Figures 3.1a and 3.1b show typical arrangements for testing slabs and beams, and columns. Typical dimensions of a standard fire test furnace are 4 m horizontally and 3 m vertically.

The assessment of standard fire resistance testing is according to load bearing, insulation and integrity. Insulation is concerned with excessive temperature increase on the unexposed surface of the test specimen. Integrity failure is associated with fire spread through gaps in the test specimen. These three failure criteria are sketched in Figure 3.2. For a load bearing steel or composite member, the load bearing requirement often governs. Load bearing failure is deemed to have occurred when the test specimen fails to support the test load or additionally for a horizontal specimen:

- the maximum deflection exceeds $L/20$, where L is the span of the specimen (in mm); or
- the rate of deflection (in mm/min) exceeds $L^2/(9000d)$, where d is the depth of the test specimen (in mm).

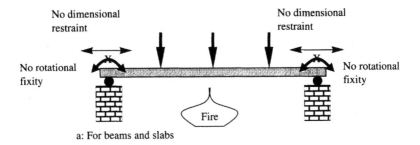

No dimensional restraint

No dimensional restraint

No rotational fixity

No rotational fixity

Fire

a: For beams and slabs

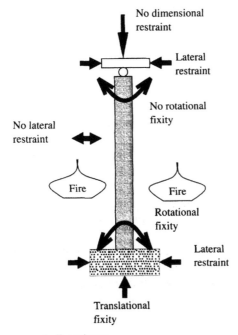

No dimensional restraint

Lateral restraint

No rotational fixity

No lateral restraint

Fire

Fire

Rotational fixity

Lateral restraint

Translational fixity

b: For columns

Figure 3.1 Typical arrangements of structural elements in standard fire resistance tests. Reproduced with the permission of the British Standards Institution under licence number 2001SK/0298.

3.2.2 A critical assessment of the standard fire resistance test method

Although the standard fire resistance test is a convenient way for quality control and grading the relative fire performance of different types of structural members, for a number of reasons, it is not very effective in developing our understanding of realistic structural behaviour in fire.

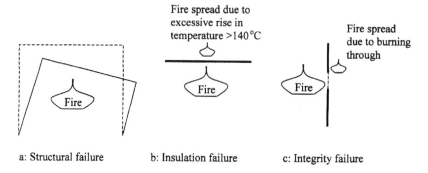

Fire spread due to excessive rise in temperature >140 °C

Fire spread due to burning through

Fire

Fire

Fire

a: Structural failure   b: Insulation failure   c: Integrity failure

Figure 3.2 Three failure modes of standard fire resistance tests.

The following list gives some of the main deficiencies of the standard fire resistance test method.

- The standard fire exposure is only one of numerous types of realistic fire conditions.
- Standard fire resistance tests are carried out on individual structural elements, not structural assemblies. Therefore, structural interactions cannot be assessed.
- Standard fire resistance tests are carried out for very specific objectives and instrumentation is usually not adequate for thorough retrospective analyses.
- A standard fire resistance test furnace is usually constructed in a commercial fire testing station for repetitive testing of construction elements as part of product development and obtaining an accredited certificate by the product manufacturer. Because of the need for the fire testing station to carry out fire tests quickly, the standard fire resistance test furnace usually has fixed dimensions and fixed controls for minimum set-up time. This often limits the dimensions of a structural element that can be tested and therefore only covers a narrow range of structural behaviour.
- The boundary condition of the structural element under testing is usually intended to be simply supported. However, it is inevitable that the real test restraint would be different. Any stiffness of this inevitable restraint could have a significant influence on the structural behaviour. However, due to inadequate instrumentation, it is often very difficult to determine the actual boundary condition in retrospective analyses. In any case, the standard fire resistance test can simulate only a very limited range of support conditions.
- The failure criteria usually do not adequately describe the intended real use.

Despite all these shortcomings, the collective results of different standard fire resistance tests have made great contributions to our understanding of the behaviour of steel structural elements in fire. It is based on these results that fire tests on complete structures under realistic fire conditions have been made possible.

3.3   FIRE TESTS ON STEEL COLUMNS

Columns are under predominantly axial compression. The three modes of failure of a steel column are local buckling, global buckling and yielding:

1   Local buckling occurs when the width to thickness ratio of the column plate is very large. Local buckling usually occurs in cold-formed thin-walled steel sections. Hot rolled steel sections have low width to thickness ratios and local buckling is very rare.
2   Global buckling is associated with the behaviour of a long column and involves lateral movement of the column along its entire length. Global buckling is associated with a lack of stiffness and can occur with or without local buckling.
3   Yielding describes the situation when all fibres in the column cross-section have reached their yield stress. This can only occur when both local and global buckling are eliminated.

3.3.1   Cross-section yield

If local buckling does not occur, complete yielding of steel in compression can only occur in short columns with a length to width ratio of not exceeding about 5 (or slenderness of about 20). The standard fire resistance test furnace is usually about 3 m high and even the largest hot-rolled column section would give a slenderness ratio of about 30. This implies that it is rare for a steel column to reach complete yield in the standard fire resistance test. On the other hand, mechanical testing of steel coupons are usually carried out in tension. However, it is often assumed that steel has the same behaviour in either tension or compression so that the tensile coupon test results can also be used to describe complete yielding of steel in compression. This assumption has been extended from ambient temperature to elevated temperatures. The results of numerous analyses and calculations that incorporate this assumption do not seem to suggest otherwise.

3.3.2   Global buckling behaviour

Figure 3.3 shows the typical axial deformation–temperature relationship of an axially loaded steel column with uniform heating. It may be divided into

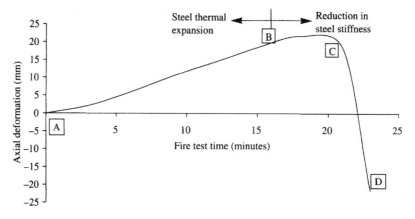

Figure 3.3 Typical behaviour of a steel column exposed to the standard fire on all sides (from Wainman and Kirby 1987). Reproduced with the permission of Corus.

three stages: the first stage (A–B) is essentially due to free thermal expansion. At high steel temperatures (B–C), the rate of increase in the column axial deformation is reduced when the column stiffness is reduced and the mechanical shortening becomes important. Finally (C–D), the mechanical shortening overtakes the free thermal expansion of the column. The column axial deformation changes direction and the column starts to contract until the column cannot sustain the applied load. The column mechanical shortening is directly related to the tangent stiffness of the column at elevated temperatures. Since the tangent stiffness reduces rapidly, the final stage is short.

Over many years, a large number of standard fire resistance tests on steel columns have been carried out. Corus (formerly British Steel) compiled a compendium containing a large number of the UK's standard fire resistance test results on steel structures (Wainman and Kirby 1987; 1988). Franssen *et al.* also reported a database containing many column fire tests (Franssen *et al.* 1998; Talamona *et al.* 1996). It is worth discussing the few issues raised by Franssen *et al.*

3.3.2.1 Boundary conditions

Under global buckling, the effective length of a column is the most important parameter that should be accurately determined. The effective length is highly dependent on the boundary condition of the column. Unfortunately, boundary conditions of many standard fire resistance tests seem to be ill defined. Often, the loading platen was in direct contact with the column head. As shown in Figure 3.4a, if the column was perfectly loaded

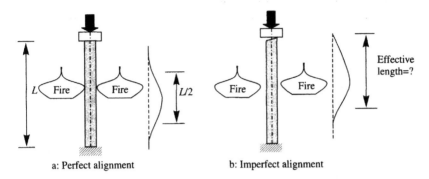

a: Perfect alignment   b: Imperfect alignment

Figure 3.4 Possible boundary conditions of a column in standard fire resistance tests.

through its centre and the column was initially perfect, the column ends may be assumed to be rotationally fixed with the column effective length being 1/2 times its physical length. However, with imperfections in the column (Figure 3.4b), any slight tendency for the column to rotate at the ends would affect the column boundary condition. Under this circumstance, analysis of the test results should take careful consideration of the column support condition. The database of Franssen *et al.* indicates that the tests of Aribert and Randrianstara (1980) and Azpiau and Unanue (1993) may be considered to have well-defined boundary conditions. In these tests loads were applied either through a very sharp knife support or a roller.

3.3.2.2 *Temperature distributions*

Temperature distribution in a test column is an important parameter. In the standard fire resistance test, the combustible volatiles are well mixed and if the column is exposed to fire attack on all sides, it can be expected that the combustion gas temperature will be nearly uniform. However, in real fire conditions, there is often a great degree of non-uniformity in the fire temperature, which will almost certainly result in non-uniform temperature distribution in the steel column, both along the column length and across its cross-section. Numerical analysis involving non-uniform temperature distributions does not present any great difficulty. However, they should be calibrated against test results where non-uniform temperature distributions have been encountered. The UK compendium of standard fire test data (Wainman and Kirby 1987, 1988) includes non-uniform temperature distributions in columns which had blocked-in webs or were partially encased in walls. Tests by Kruppa (1981–82) on external columns outside the fire compartment give some information of non-uniform temperature distributions in all directions. Non-uniform temperature distributions in steel columns were also observed and recorded in tests by Aasen (1985).

3.3.2.3 *Loading conditions*

All reported column fire tests were carried out under transient state testing, i.e. the load was applied on the test column first. The applied load was usually the working load, representing about 60% of the ultimate strength of the column at ambient temperature. The load was kept constant when the column was exposed to fire attack. The test was terminated when the applied load could not be maintained. Therefore, from these tests, it is not possible to study the column behaviour under increasing temperature but reducing loads. Such loading conditions can become possible when a column forms part of a complete structure and sheds its load onto the adjacent structure at increasing temperatures.

3.3.3   Local buckling

There are very few reported fire tests in the literature that are concerned with the behaviour of thin-walled steel columns in fire, where local buckling is most likely.

Gerlich (1995) reported the results of three fire tests on wall panels constructed using thin-walled steel studs, two with lipped channel sections of $102 \times 51 \times 1.0$ mm and one with unlipped channel sections of $76 \times 32 \times 1.15$ mm. The wall panel was 2850 mm high for the unlipped channels and 3600 mm for the lipped channels. Gypsum plasterboard linings were attached to both flanges of the channels using screws spaced at 300 mm centres. Fire exposure was on one side and was according to ISO 834 (ISO 1975). Test results show substantial temperature gradients in the steel studs such that lateral deflection was observed from the beginning of the fire test due to thermal bowing. Local buckling of the steel studs was also observed. The failure modes of the three different wall panels suggest that structural contributions from the "non-structural" plasterboard linings should be taken into consideration. In two tests where the plasterboard linings stayed intact, global buckling of the steel studs about the minor axis was prevented and the steel studs were forced to buckle about the major axis. When the lining material was destroyed during one test, the failure mode was flexural-torsional buckling.

Ala-Outinen and Myllymaki (1995) reported the results of some local buckling tests on thin-walled rectangular steel tubes at elevated temperatures. Tests were carried out under steady-state condition, i.e. the steel temperature was raised to the specified target values and the specimen was then loaded to failure.

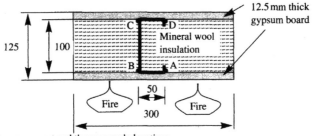

a: Test arrangement and thermocouple locations

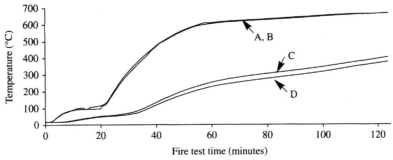

b: Recorded temperature–time relationships

Figure 3.5 Temperature distribution in a panel exposed to fire on one side (from Feng et al. 2001).

Feng *et al.* (2001) reported some results of fire tests on unloaded small panels (300 mm by 300 mm) constructed using gypsum boards, mineral wool insulation and channel studs. The test arrangement is sketched in Figure 3.5a. Figure 3.5b presents measured temperatures in the steel channel at a few representative locations from one test. Clearly, because fire exposure was from one side, temperature gradients in the steel channel were high. There was also some temperature gradient along the flange on the unexposed side. This is due to the fact that the steel web conducts heat much more rapidly than the mineral wool insulation. Sultan (1996) also reported some results of non-uniform temperature distributions.

From the same series of tests, Feng *et al.* also reported some elevated temperature load tests on short channel sections. These tests were carried out under steady state in an electrically heated kiln. Recorded temperatures indicate near uniform temperature distribution in the column. The target temperatures were 250 °C, 400 °C, 550 °C and 700 °C. Figure 3.6 shows a group of samples after test. It is clear that thin-walled columns undergo a variety of buckling modes, including local buckling, distortional buckling, global flexural buckling and torsional buckling.

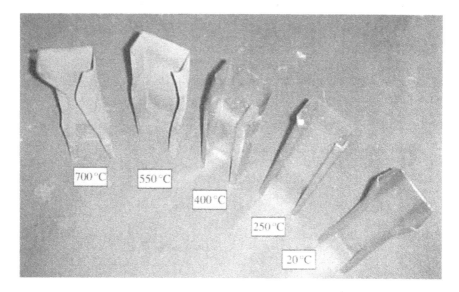

Figure 3.6 Deformed shape of thin-walled stub columns at different temperatures (from Feng *et al.* 2001).

Alfawakhiri and Sultan (2000) reported the results of six fire tests on axially loaded lightweight steel framed (LSF) assemblies exposed to fire attack on one side. The assemblies were constructed of thin-walled lipped channels and protected by two layers of gypsum boards on each side. Four of these assemblies incorporated interior insulation and the other two had no interior insulation. During the early stage of the fire tests, the LSF assemblies bowed towards the fire test furnace due to thermal bowing caused by the temperature gradients in the steel channels. During the later stage of the fire tests, the LSF assemblies without interior insulation continued to bow towards the furnace and the assemblies eventually failed by compressive crushing on the cold face in the middle of the test assemblies. However, deformations in the LSF assemblies with interior insulation reversed in direction and the assemblies failed by compressive crushing on the hot face at the location of the service hole near one end of the assemblies. This difference in behaviour can be explained by the difference in temperature gradients in the interiorly insulated and uninsulated assemblies. In the interiorly uninsulated ones, the temperature gradients became small and the thermal bowing effect dominated the structural behaviour. In the interiorly insulated ones, the temperature gradients continued to be high and caused a large shift of the centroid of the steel channel section towards the cold side. This caused compression on the hot side and dominated the structural behaviour during the later stage of heating. The results of these fire tests suggest that interior insulation caused a reduction in the fire resistance of loadbearing LSF walls.

3.3.4 Summary of fire tests on isolated steel columns

From this short review, the following observations may be made.

1   For hot-rolled steel columns:

- failure modes of a column in fire are the same as those at ambient temperature;
- sufficient test results are available to develop complete understanding of the column behaviour in fire;
- when analyzing the results of a fire test, it is important to pay attention to the support condition and non-uniform temperature distribution in the column; and
- some additional tests may be necessary to understand the unloading behaviour of a column under increasing temperatures.

2   For cold-formed thin-walled columns:

- temperature gradients can be high due to fire exposure on one side;
- it is important to consider the effects of thermal bowing and shift of the centroid of cross-sections, both caused by temperature gradients;
- it is necessary to consider the effects of service holes in steel members;
- it is important to consider contributions of the non-structural components;
- there are some experimental information to study local buckling when the temperature distribution is uniform;
- no information is available to help understand local buckling when the temperature distribution is non-uniform; and
- very little information is available on interactions of different buckling modes in fire.

3.4 FIRE TESTS ON RESTRAINED COLUMNS

When a heated steel column forms part of a structure in fire, it will be subject to different types of restraint by the adjacent structure. As a result, loads in the column can change during the fire exposure. The column axial load changes due to restrained thermal expansion. Bending moments in the column are affected by the variable column bending stiffness relative to the adjacent structure and by the P-δ effect induced by column thermal bowing and displacements of the adjacent beams.

3.4.1 Effects of restrained thermal expansion

Simms *et al.* (1995–96) experimentally studied the changes in axial loads and failure temperatures of a centrally loaded steel column with different degrees of axial restraint. The rate of temperature increase in the column was approximately 10 °C/min and near uniform temperature distribution was achieved. The axial restraint was simulated using springs and was effective only when the column was expanding and was not effective when the column was contracting. From their test results, they observed increases in the column axial load, and the higher the restraint stiffness, the higher the increase in the column axial load. The test column failed rapidly when the maximum axial load reached the column buckling capacity at the elevated temperature. Figures 3.7a and 3.7b plot typical relationships of the column axial load and lateral deflection against temperature. The reduction in the column restraint force approaching failure is clearly due to the column shortening as a result of accelerating lateral deflection of the column.

Correia Rodrigues *et al.* (2000) carried out similar experimental studies. However, there were some notable differences:

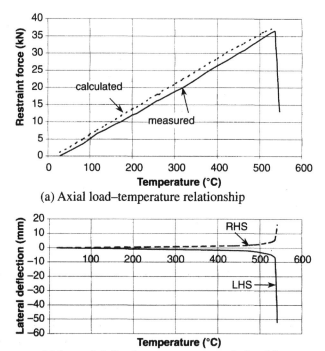

(a) Axial load–temperature relationship

(b) Lateral deflection–temperature relationships

Figure 3.7 Typical behaviour of an axially restrained column (from Simms *et al.* 1995–96).

- The tests of Simms *et al.* were on columns using realistic cross-sections and those by Correia Rodrigues *et al.* used short bars of small rectangular cross-sections of high slenderness.
- Simms *et al.* used a propane gas burner and achieved almost uniform temperature distribution in the column. Correia Rodrigues *et al.* used an electrically heated kiln and recorded significant temperature variation in the longitudinal direction of the column.
- All columns were concentrically loaded in the tests of Simms *et al.* and there were some eccentrically loaded columns in those of Correia Rodrigues *et al.*
- The most important difference in these two studies was the restraint stiffness. In the tests of Simms *et al.*, the axial restraint was only effective when the column was expanding. In the tests of Correia Rodrigues *et al.*, the restraint stiffness was available during the entire test period, i.e. when the column was either expanding or contracting.

The test results of Correia Rodrigues *et al.* provided some interesting observations. Figures 3.8 shows typical recorded axial load–column temperature relationships for a concentrically and an eccentrically loaded column. Consider the concentrically loaded column. If the restraint stiffness was sufficiently high, the column unloading was initially sharp after the total axial load in the column had exceeded its buckling capacity at the elevated temperature. However, the column was able to find a stable position, after which, column unloading was gradual. As will be discussed later (in Section 5.3.2), the column behaviour after attaining the maximum load can make a substantial difference to the column survival time in fire. For an eccentrically loaded column, the column unloading was gradual at all restraint stiffness.

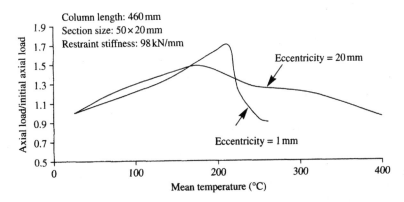

Figure 3.8 Behaviour of an axially restrained column in fire (Reprinted from the *Fire Safety Journal*, "Experimental research on the critical temperature of compressed steel elements with restrained thermal elongation". Vol. **35**, pp. 77–98, Correia Rodrigues *et al.* (2000) with permission from Elsevier Science).

Lennon and Simms (1993) and Lennon (1994) reported some fire tests on restrained columns. However, these tests were carried out in a real structure (the Cardington test frame). Since the structure used nominally simple connections, the restraint stiffness to the test columns was small and increases in column compressive loads were low.

3.4.2 Effects of rotational restraints

Rotational restraint to a column can affect its behaviour in two ways: (1) to change its effective length; and (2) its bending moments. No fire test results could be found that concentrated on column effective length and tests to study changes in column bending moments will be described in Section 3.9.

3.4.3 Summary of fire tests on restrained columns

Although a column will inevitably be restrained by the adjacent structure when forming part of a structure, it is apparent that only a few fire tests have addressed this issue. From observations of these tests, it can be seen that:

- axial restraint to a column will increase the compressive load in the column;
- the restraint has a dual function in increasing the column compressive load when the column is expanding and in stabilizing the column after buckling and when the column is contracting; and
- test information is required to quantify the effect of rotational restraint on the slenderness of a column.

3.5 FIRE TESTS ON COMPOSITE COLUMNS

3.5.1 Local buckling of steel

When steel is restrained by concrete, as in a composite column, local buckling of the steel is much less likely than in a bare steel column. In fact, if the steel is encased in concrete, local buckling will not occur. Local buckling may occur in concrete filled columns. Depending on the type of concrete filled columns, local buckling may or may not be important. If the column is simply supported at its ends, local buckling is unlikely to be important. However, if the column forms part of a complete structure, as will be discussed in Section 5.3.4, the effective length of the column may differ depending on where local buckling is located. Although local buckling has been observed in fire tests, reporting on this phenomenon has never been detailed and its occurrence appears almost to be random (Edwards 1998a), depending on such factors as temperature distributions in the column and local quality of concrete.

3.5.2   Global buckling behaviour

3.5.2.1   Composite columns with steel encased in concrete

In a contract report to the Commission of European Community (CEC 1987), the results of a number of fire tests on composite columns made of steel sections with concrete encasement in between the flanges are described. The behaviour of this type of columns is very similar to that of bare steel columns. A problem of this type of column is spalling of the concrete. Fortunately, in one test where spalling of the concrete was observed, the reinforcement was not exposed, thus the test column still achieved high fire resistance. Nevertheless, this does suggest that this type of construction may not be suitable to concrete that is suspect to spalling under fire conditions, such as plain high strength concrete.

3.5.2.2   Unprotected concrete filled tubes

Over a number of years, the National Research Council of Canada (NRCC) carried out numerous fire tests on unprotected concrete filled steel columns. The fire tests were carried out in a specially constructed fire testing furnace at the National Fire Laboratory (Lie 1980). The Canadian fire tests included different dimensions and thickness of circular and square hollow sections, a variety of concretes (plain, bar reinforced, high strength and steel fibre reinforced, with either siliceous or carbonate aggregates) and different levels of applied loads (Kodur 1998; Lie and Chabot 1992; Lie and Kodur 1996). All the test columns were 3810 mm and the column ends were rotationally restrained at the ends. All fire tests were carried out with unprotected steel sections and the fire exposure was according to ASTM E-119 (ASTM 1985).

Figure 3.9 shows a typical response of the recorded axial deformation–time relationship. It can be divided into four parts: (1) a phase of steady increase in the column expansion (A–B) is followed by (2) a sharp contraction (B–C) and (3) then gradual contraction (C–D) in the column axial deformation; and (4) the column experiences another sharp contraction (D–E) before failure.

This type of behaviour may be explained by considering the temperatures and resistances of the steel tube and the concrete. Before heating starts, the steel tube and the concrete core share the applied load in composite action. During the early stages of the fire test, because the steel is at a much higher temperature, it expands faster than the concrete. The applied load is now mainly resisted by the steel tube and the first phase corresponds to thermal expansion of the steel. As the steel temperature increases, it loses its load carrying capacity and the column suddenly contracts due to buckling of the steel tube. This is reflected in the second stage behaviour and is often

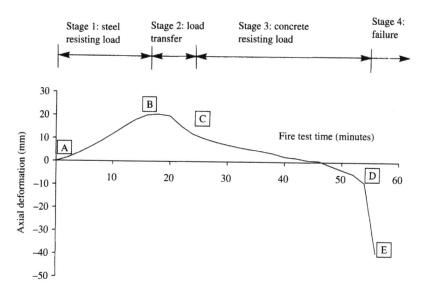

Figure 3.9 Typical time–axial deformation response of a concrete filled column (from Lie and Chabot 1992).

accompanied by local bulging of the steel tube in a fire test. If the concrete core has sufficient load carrying capacity, the applied load will be shed from the steel tube to the concrete core when the steel tube has contracted in length to the level of the concrete core. The column response is now characterized by a gradual contraction until the applied load exceeds the combined resistance of the steel tube and the concrete core at much higher temperatures. Since the load is mainly resisted by the concrete core, the thickness of the steel tube has very little influence on the fire resistance time of the composite column.

Fire tests by others (CEC 1987; Sakumoto *et al.* 1993) on unprotected concrete filled columns show similar results.

3.5.2.3 *Protected concrete filled columns*

Unprotected composite columns have inherently high fire resistance. Nevertheless, it is sometimes necessary to improve the fire resistance of concrete filled columns using external fire protection. Compared to unprotected columns, only a few fire tests on protected columns have been reported. Among these, Edwards (1998b, 2001) reported the results of six fire tests on protected concrete filled columns. Sakumoto *et al.* (1993)

reported a few fire tests on protected columns. An interesting feature of the tests of Sakumoto *et al.* is that the so-called FR steel (see Chapter 5) was used. The difference between conventional steel and FR steel was mainly in the strength of the steels at temperatures of around 600 °C (cf. Figure 5.4).

For concentrically loaded columns, the observed deformation behaviour of the protected concrete filled columns of Edwards and Sakumoto *et al.* was different. Whilst the behaviour of Sakumoto *et al.* was similar to that of unprotected columns, the behaviour observed by Edwards was different. In the tests described by Edwards, the column behaviour was similar to that of a steel column, as shown in Figure 3.3. This is an indication that due to external fire protection, the steel and concrete temperatures were similar and the applied load was shared between the steel and the concrete during the entire course of the fire test. On transfer of load from the steel tube to the concrete core, failure was rapid due to inability of the hot concrete core alone to resist the applied load.

It is interesting to compare the effects of steel thickness between the NRCC (Lie and Chabot 1992) fire tests where unprotected steel tubes were used and the tests of Sakumoto *et al.* where the steel tubes were protected. In both cases, the applied load was first resisted by the steel tube then transferred to the concrete core after buckling of the steel tube. Whilst the results of NRCC tests indicate that the steel tube thickness had minimal influence on the load carrying capacity and fire resistance of the composite column, the tests of Sakumoto *et al.* suggest that there was some post-buckling strength from the fire exposed steel tube to contribute to the resistance of the composite column. However, post-buckling response and strength of steel tubes in composite columns at elevated temperatures have not been experimentally investigated in any detail.

3.5.2.4   High strength concrete (HSC) filled columns

One of the main advantages of concrete filled columns is the high strength and stiffness gained from using small cross-sections. It is natural to consider using high strength concrete to further improve the structural efficiency of this type of construction. High strength concrete refers to concrete that has a cylinder strength of higher than about 60 N/mm^2. However, high strength concrete (HSC) is different from normal strength concrete (NSC) in two important aspects that affect fire performance. The water content in HSC is lower than in NSC, resulting in higher temperatures in HSC filled columns. At elevated temperatures, the strength and stiffness of HSC are lower than those of NSC. Both properties of HSC lead to lower fire resistance of HSC filled columns compared to NSC filled ones. For example, Figure 3.10 compares the time–axial deformation relationships of a HSC and NSC filled column. both columns had similar level of load ratio. The fire tests were

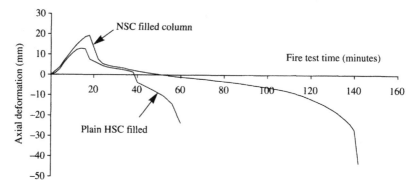

Figure 3.10 Comparison between NSC and plain HSC filled columns (from Kodur and Wang 2001).

carried out at the National Research Council of Canada (Kodur 1998, Kodur and Wang 2001). From Figure 3.10, it can be observed that the fire resistance of the HSC filled column was much shorter than that of the NSC filled column. In particular, the third phase of the column behaviour shown in Figure 3.9 (when the applied load was resisted by concrete) was very short. This is mainly due to rapid reduction in the strength of HSC at high temperatures.

A small amount of high strength steel fibres may be added in HSC to improve the fire performance of HSC filled columns (Kodur 1998). The corrugated shape of these steel fibres provides a strong mechanical bond to the concrete. This increases the concrete temperature at which the HSC strength starts to decrease from about 200 °C to about 500 °C. For example, Figure 3.11 compares the measured time–axial deformation relationships of a NSC filled column and a fibre reinforced HSC filled column. The behaviour of the HSC filled column is similar to that of the NSC filled column, with the fibre reinforced HSC being able to resist the applied load for a long period of time. For this column, the percentage of steel fibres in the concrete mix was 1.77% by weight.

In concrete filled columns, the concrete is inside the steel tube so that the problem of high strength concrete spalling is eliminated (Hass *et al.* 2001).

3.5.3 Restrained composite columns

The restraining factors that affect the behaviour of a steel column that forms part of a complete structure are equally applicable to a composite column. However, there is no reported fire test in the literature on restrained composite columns. From the previously described experimental observations,

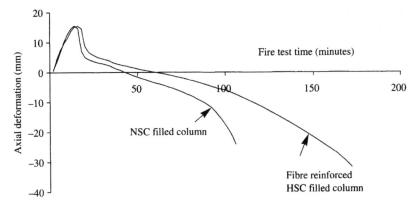

Figure 3.11 Comparison between NSC and fibre reinforced HSC filled columns (from Kodur and Wang 2001).

either the steel or concrete in a composite column can resist the applied load. This makes the behaviour of a restrained composite column in fire much more complicated than that of a restrained steel column. For example, considering the behaviour of an axially restrained composite column, depending on whether the steel or the concrete is resisting the applied load, the stiffness of the composite column will be different and different additional compressive load will be induced in the column. Coupling this with the uncertainty on the post-buckling strength of the steel in a composite column, detailed investigation in this area is obviously necessary.

3.5.4 Summary of fire tests on composite columns

Composite columns have inherently higher fire resistance than steel columns, yet the behaviour of composite columns in fire has been studied in a much less detailed way. From available experimental observations, the following conclusions may be drawn.

- The behaviour of a composite column with encased steel section is similar to that of a bare steel column, except there is no local buckling of the steel section. However, the effect of concrete spalling should be considered.
- The behaviour of a concrete filled composite column is more complicated. The steel tube and the concrete core resist the applied load at different stages of fire exposure, which makes it doubtful whether composite action can still be maintained. This will depend on the post-

buckling strength of the steel tube, for which there does not appear to exist detailed experimental investigations.

- Only a very limited improvement in fire performance can be obtained by using plain HSC. Much greater improvement can be achieved by using a small amount of steel fibres in the HSC concrete mix.
- At ambient temperature, if a concrete filled circular hollow section is short (height/diameter about 3), confinement of the concrete may be used to enhance the column squash load. No test result has been found which addresses this issue in fire. However, it is unlikely that confinement will be high in fire when the steel is at a higher temperature and expands more than the concrete core.
- No experimental study on the behaviour of restrained composite columns has been found.

3.6 FIRE TESTS ON STEEL AND COMPOSITE BEAMS

The behaviour of a steel beam is much more complicated than that of a steel column. It can undergo local buckling, cross-sectional yielding, LTB and shear buckling. As will be discussed later (Section 5.5.2), if the beam's deflection is large and the ends of the beam are longitudinally restrained, catenary action may also develop in the beam. Moreover, a beam usually has non-uniform temperature distribution in the cross-section. The interaction of this with the different modes of behaviour makes the behaviour of a beam in fire extremely complex.

3.6.1 Behaviour of bare steel beams in fire

3.6.1.1 Local buckling

In the most common type of beam construction where the compression (top) flange of the beam is restrained, e.g. by concrete floor slabs, and the unrestrained bottom flange is in tension, local buckling is only possible in the web. When under bending stresses, the stress distribution in the web is most favourable for resisting local buckling. Therefore, local buckling of the web can only occur when the depth over thickness ratio of the web is very high. In hot-rolled steel sections, local buckling of the web is rare. Local buckling is most likely to occur in cold-formed thin-walled steel beams. Compared to the problem of local buckling in a steel column, the beam problem is much more complex, involving non-uniform stress and temperature distributions. However, there is a clear lack of test information in this area.

3.6.1.2 Cross-sectional yield

Standard fire resistance tests are usually conducted on simply supported steel beams with floor slabs on top. Under this condition, local buckling and LTB are unlikely to occur. The bending behaviour of the beam is governed by cross-sectional yielding of steel.

The most comprehensive test information on the bending behaviour of steel beams in fire is provided by the two volume compendium of the UK's standard fire resistance test data (Wainman and Kirby 1987, 1988). Figure 3.12 shows the typical deflection–time relationship of a simply supported beam. The behaviour is relatively simple. The deflection of the beam increases at an almost constant rate, until about 20 min into the fire test when the rate of the beam deflection accelerated rapidly. Failure of the beam follows quickly at about 25 min into the fire test with very large deflections in the beam, often referred to as "runaway".

The lateral deflection of the beam in the early stage of the fire test is primarily due to thermal bowing as a result of temperature gradient in the cross-section of the beam. Acceleration of the beam deflection is due to rapid reduction in the beam's strength and stiffness at higher temperatures. Bending failure occurs when the maximum bending moment in the beam has attained the plastic bending moment capacity of the cross-section at elevated temperatures.

3.6.1.3 Lateral torsional bucking

Piloto and Vila Real (2000) appear to be the only one to have carried out fire tests to study LTB of steel beams at elevated temperatures. Tests

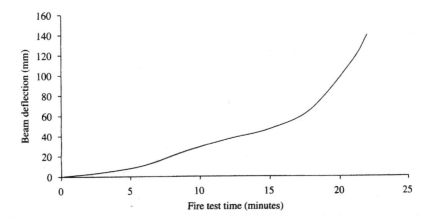

Figure 3.12 Typical behaviour of a simply supported steel beam under standard fire exposure (from Wainman and Kirby 1987). Reproduced with the permission of Corus.

were carried out under the steady-state condition on a 3.5 m long beam made of an IPE 100 section. Heating was by electrical blankets. As part of this investigation, they also measured residual stresses and initial deflections of the test beams. Because the test beam was uniformly heated, they did not report any special incidents in addition to those at ambient temperature.

3.6.2   Composite beams

A large number of fire tests have been carried out on composite beams. They include:

- tests on unprotected, simply supported conventional composite beams (Wainman and Kirby 1987, 1988; Zhao and Kruppa 1995);
- tests on protected simply supported conventional composite beams (Newman and Lawson 1991); and
- tests on partially encased steel section and partially encased steel section in composite action with concrete floor slabs (CEC 1987; Hosser *et al.* 1994; Kordina 1989).

Except for the tests of Zhao and Kruppa (1995), results of all other tests indicate that the structural performance of a composite beam is similar to that of a steel beam. Failure of a composite beam was by tensile yielding of the steel section, implying that the plastic bending moment capacity of the composite cross-section was reached. Moreover, shear connectors in these tests performed well and were not the weakest link.

In the tests reported by Newman and Lawson (1991), the steel beam was fire protected using different fire protection systems. One objective of these tests was to compare the performance of composite beams with filled and unfilled voids between the protected steel beam and the profiled steel deck (see Figure 3.13). They observed that the effect of unfilled voids was to increase the temperature of the upper flange. Typically, if the voids were filled, the upper flange temperature was about 20% lower than the lower

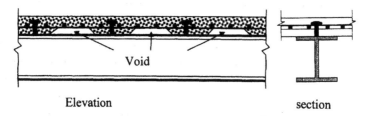

Elevation   section

Figure 3.13 Cross-sections of a structural floor system.

flange temperature. In the case of unfilled voids, the upper flange temperature was higher than the lower flange temperature, the difference depending on the severity of fire exposure. Typically, the upper flange temperatures were about 10% and 30% higher than the lower flange temperature at 60 and 90 min of the standard fire exposure respectively.

The tests of Zhao and Kruppa (1995) were different from the others. In addition to simply supported beams they also tested continuous beams. In these tests, measurements were taken of the slip between the steel beam and the concrete flange at the ends of the test beam. As part of this investigation, they also reported some load–slip relationships for shear connectors at high temperatures.

The main observations from the Zhao and Kruppa fire tests are listed as:

- When the steel section was unprotected, failure of the composite beam was due to tensile yielding of the steel section as in other tests.
- When using a protected steel profile and partial shear connection, they observed significant slip (>10 mm) between the concrete slab and the steel profile at the ends of the beam due to the fact that the protected steel profile was at a much lower temperature than the concrete slab at the shear connector interface. This large slip may cause shear connector failure, leading to failure of the composite beam.
- When testing continuous beams with a cantilever span, they observed local buckling in the web and lower flange of the steel section at the middle support.
- When testing continuous beams, they observed local buckling and bearing failure in the web near the centre support where there was a concentrated load. At the same location, they also observed fracture of shear connectors.
- Temperatures in the steel section near the supports were much lower than in the span. In a continuous beam where the hogging bending moment capacity governs design, the beam's strength may be much higher than that assuming uniform temperature distribution in the beam.

3.6.3 Restrained beams

If a beam forms part of a complete structure, the beam will interact with the adjacent structure. Both support and loading conditions of the beam may change as a result of these interactions. At present, not enough experimental information is available to enable reliable analysis of a restrained beam. A particularly useful exploitation of the restraining effect on a beam is to make use of catenary action. Catenary action in a beam occurs when the beam is restrained in length at its ends and it undergoes large lateral deflections. The applied lateral load in the beam is resisted by the vertical components of the catenary force in the beam. To make use of catenary action, studies should be carried out to investigate how the catenary force in the beam may be anchored at the beam ends and on how the various modes of bending behaviour, i.e. local buckling, cross-sectional yielding and LTB may affect, or be affected by, the development of this load carrying mechanism. Sections 5.5.2 and 5.5.3 will give a more detailed discussion of this topic.

3.6.4 Summary of fire tests on beams

The following modes of behaviour may be expected:

- if the beam is simply supported, failure of the beam is governed by tensile yielding of the steel section;
- if concentrated loads exist, web buckling and bearing failure should be considered as at ambient temperature;
- LTB in the hogging region near interior supports may occur;
- shear connectors are usually not the weak link of a composite beam. However, shear connector failure may occur in some cases and experiments are necessary to establish load–slip relationships of shear connectors at elevated temperatures; and
- future fire tests are necessary to understand the behaviour of restrained beams.

3.7 FIRE TESTS ON SLABS

Composite construction consisting of concrete cast *in situ* on top of the steel decking is conventionally used in construction as floor slabs. To satisfy the requirements of fire regulations, the manufacturers have carried out many standard fire resistance tests. A review of these fire tests is given in a paper by Cooke *et al.* (1988). An analysis of the results of fire tests by Cooke *et al.* suggests that structural failure of a composite slab is due to the formation of a plastic hinge mechanism, therefore, the plastic design method may be used to evaluate the load carrying capacity of the composite slab.

Cooke *et al.* also reported some results of temperature distribution in composite slabs above steel sections. Due to the shielding effect of the steel sections, temperatures in a composite slab near the steel sections were substantially lower than those away from the steel sections. Considering a uniformly loaded continuous slab as shown in Figure 3.14, the hogging bending moment capacity usually governs design. Since the hogging bending moment reduces sharply across the width of the supporting sections, benefit could be obtained if consideration is given to this favourable distribution of

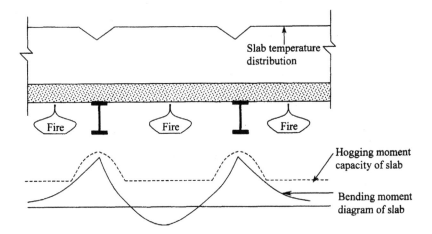

Figure 3.14 Effect of non-uniform longitudinal temperature distribution on the fire resistance of a continuous slab.

temperatures in the composite slab near the steel sections. Under this circumstance, it is only necessary to ensure that the hogging bending moment capacity of the slab in the span is not less than the much reduced hogging bending moment away from the supports.

3.8   FIRE TESTS ON CONNECTIONS

Whilst numerous experimental studies have been performed to evaluate the performance of connections at ambient temperature (Nethercot 1985), studies on connection behaviour under fire conditions are relatively recent and few. Since connections are usually designed to be pin-joints, but have some bending moment resistance, the aim of these few fire investigations was primarily to address the increased fire resistance for the connected steel beam. They also follow the tradition of testing isolated connections at ambient temperatures. Although experiences from frame connections (Sections 3.10.2.2) cast doubt on the suitability of using isolated connections to represent frame connections in fire, it is nevertheless useful to review the isolated connection tests so as to have some reference information on which to build understanding of frame connections in fire.

3.8.1   SCI tests

Results of early fire tests on connection behaviour have been reported by Lawson (1990a,b). In this series of connection tests, eight pairs of bolted beam to column connections were made in a cruciform arrangement and subjected to the standard fire exposure. These eight connections included four flush end plate connections (two using steel beams, one composite beam and one shelf-angle beam), two extended end plate connections (both using steel beams) and two web cleat connections (one using steel and one composite beam). The ends of the beams were free to expand and contract. In the test, the connection bending moment was fixed at about 1/3 or 2/3 of the connection bending moment capacity calculated at ambient temperature. Measurements taken include temperatures in various parts of the connection and increasing rotation of the connection at elevated temperatures. It was found that temperatures in the connection region were much lower than that of the lower flange of the steel beam, which is normally used to assess the beam behaviour. Temperatures in the exposed bolts were about 100–150 °C lower and those inside the concrete slab about 300–350 °C lower than the temperature of the lower flange of the steel section. Failure of the connection was due to plastic deformations of the plates. Bolts were not the weak link of the connection.

3.8.2 Collaborative investigations between the University of Sheffield and Building Research Establishment

Leston-Jones (1997) and Leston-Jones *et al.* (1997) reported the results of some fire tests on steel and composite beam to column connections. Eight connections were tested, of which five were bare steel connections and three were composite connections, all using flush end plates. The cruciform arrangement shown in Figure 3.15 was adopted in these tests. Fire exposure was by wrapping a barrel furnace around the connection. Temperatures, displacements and rotations at various locations of the connection were measured to enable detailed quantification of the connection behaviour. All tests were performed by applying a fixed bending moment and then increase the furnace temperature at a rate of about 10 °C/min while maintaining the applied load.

The test results indicate that connection behaviour at elevated temperatures is similar to that at ambient temperature. For a bare steel connection, deformations were mainly concentrated in the column web in the compression zone with some contribution from the column flange in the tension zone, whilst very little deformation occurred in the beam and the end plate. For a composite connection, due to the stiffening effect of the reinforcement in the tension zone where temperature was low, deformations were mainly in the column web in the compression zone.

A comparison between the behaviour of bare steel connections and composite connections indicates that degradations in the stiffness of both types of connections were similar, but composite connections showed much less reductions in their bending moment resistance at elevated temperatures.

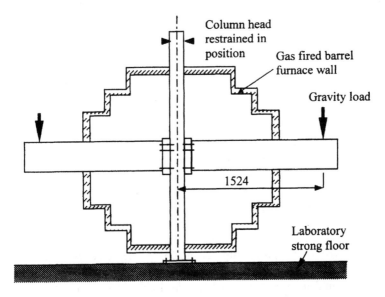

Figure 3.15 Schematic arrangement of elevated temperature connection tests (from Leston-Jones 1997).

This is to be expected since for both types of connections, the degradation in the stiffness of a connection is mainly affected by the column temperature in the compression zone, which is not affected by the connection type. On the other hand, the reinforcement in the tension zone in a composite connection is able to contribute to the bending moment resistance of the connection. Since the reinforcement temperature is low, its strength is relatively unchanged so that the relative contribution of the reinforcement becomes greater, leading to higher connection resistance and lower degradation in strength of the composite connection.

The test results were used to derive the bending moment–rotation relationships of the connections at different temperatures. Leston-Jones used the lower flange temperature of the steel beam as the reference value. There was no experimental investigation of the sensitivity of the connection moment–rotation–temperature relationships to different temperature distributions in the connection region. Nevertheless, Leston-Jones (1997) proposed a component based model to derive the connection moment–rotation–temperature relationships and achieved good agreement with the test results. The component based model may be used to deal with different temperature distributions in the connection. Indeed, further numerical analyses by Leston-Jones on the effect of connection behaviour on steel framed structures indicate that the frame behaviour is sensitive to temperature distributions in the connection region.

Al-Jabri (1999) and Al-Jabri *et al.* (1998) carried out similar studies on composite connections.

3.8.3   Behaviour of frame connections in fire

There are no specific fire tests to study the behaviour of a connection as part of a frame structure. However, the main difference between an isolated connection and a frame connection is the presence of an axial load in the frame connection. During the early stage of a fire exposure, a compressive axial load is developed in the frame connection due to restrained thermal expansion of the connected beam. This compressive load, in combination with the bending moment, can induce buckling in the connected beam near the connection. During the late stage of fire exposure when the connected beam undergoes large deflections and in catenary action or during the cooling down stage, a tensile force is developed in the frame connection. This tensile force can be high enough to fracture the connection. Therefore, the moment–rotation relationship can no longer describe the behaviour of a frame connection. A more faithful representation of the connection behaviour should include the variable axial load. At present, experimental investigations are being carried out at the University of Sheffield (Spyrou and Davison 2001) to assess whether the component method in Annex J of Eurocode 3 Part 1.1 (CEN 1994) may be modified to include the influence of axial load in a frame connection.

3.8.4   Summary of fire tests on connections

The behaviour of a connection in fire is a relatively poorly researched area. This is perhaps a consequence of the perceived satisfactory performance of connections in fire. Existing research studies on connection behaviour are intended to explore the benefits of using the bending strength of nominally pinned joints to increase the fire resistance of the connected beams. Only a few connection tests have been performed and they have concentrated on obtaining the moment–rotation relationships of isolated connections. It is doubtful whether these results will be useful when dealing with the behaviour of frame connections. An additional parameter, the axial load in connection, should be included.

3.9   FIRE TESTS ON SKELETAL FRAMES

Fire tests on statically determinate, simply supported, individual, structural elements, where the support and loading conditions are well defined, are the necessary first step towards precise quantification and understanding of the behaviour of complete structures in fire. However, they are not sufficient for an understanding of the behaviour of complete structures. Interactions between different structural elements in a complete structure can alter the loading and support conditions of any structural element. This alteration can lead to completely different structural behaviour from that based on the initial set of loading and boundary conditions. Therefore, even though tests on complete structures in fire are expensive and time consuming to perform they are an essential part of understanding structural behaviour in fire. This section describes some observations from fire tests on skeletal steel frames/ assemblies. In the next section, fire tests on complete buildings will be described.

3.9.1 Tests of Rubert and Schaumann on 1/4 scale steel frames, Germany

In this study, three different arrangements of rigidly connected 1/4, scale steel frames were tested at elevated temperatures by using electrical heating. The three test arrangements are shown in Figure 3.16. Lateral torsional buckling of the structural member was prevented by using stiffeners against torsional displacements at a number of locations. Frames EHR were braced and frames EGR and ZSR were unbraced. Rubert and Schaumann (1986) gave failure temperatures of the heated steel members. No information was provided for strains and forces attained in the test frames, these information being essential to enable quantification of the effects of interactions between different frame members. Nevertheless, these elevated temperature tests are

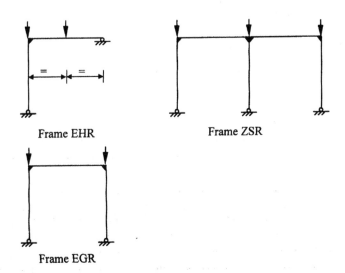

Frame EHR   Frame ZSR

Frame EGR

Figure 3.16 Test arrangements of Rubert and Schaumann (1986).

perhaps the most widely quoted tests and have been used by various researchers for the validation of their numerical models.

3.9.2 Test of Fire Research Station and Corus on a rugby post frame (Cooke and Latham 1987), UK

The Fire Research Station carried out perhaps the first fire test on a full-scale structural assembly. The test structure was a goal post assembly of a steel beam (406 × 178 × 54 UB) and two columns (203 × 203 × 52 UC). The columns were pin jointed to the test laboratory and the beam was connected to the columns using flush end plate connections. The columns had blocked-in webs. Concrete slabs were placed on top of the steel beam to give realistic heating condition to the beam, but the steel beam was separated from the concrete slabs by a layer of 25 mm thick ceramic wool blanket to prevent composite action. Bracing was provided to the test frame near the beam–column connections to prevent sway and out-of-plane deflections. The test assembly was unprotected and was enclosed within a specially built test furnace. Natural fire exposure was provided with timber cribs. Extensive measurements were made for the combustion gas temperature, the steel temperatures and deflections.

The recorded pattern of behaviour of the beam's lateral deflection was similar to that of a simply supported beam (cf. Figure 3.12). Initially, the deflection was due to thermal bowing caused by temperature gradients in the beam. In the later stages of the fire test, the beam deflection was due to increased mechanical deflection at reducing beam stiffness, leading to the runaway deflection behaviour when the reduced beam load carrying capacity could not sustain the applied load.

Due to practical difficulties, there was no measurement of strains in the structure, hence the evolution of forces in the steel framework could not be determined experimentally to study interactions in the frame. Nevertheless, this test provided a number of observations to assess differences between the behaviour of an individual element and that of a framework in fire.

Although the beam–column connections were intended to be pin-joints, there was substantial hogging bending moment transfer at the connections and plastic hinges were observed near the connections in the beam. This was confirmed by the observed twist due to LTB in the lower flange, indicating compression in this region near the supports.

The calculated equivalent time according to the measured steel temperatures gave an equivalent standard fire test time of 32 min. As discussed in Section 3.6.1, a simply supported beam would have failed at about 20 min into the standard fire resistance test. The fact that the framework survived much longer indicates substantial increase in resistance gained by connection continuity that may be exploited to provide increased fire resistance to the beam.

3.9.3   Small scale tests on rigid steel frames at Tongji University, China

Li and Jiang (1999) and Li *et al.* (1999) reported the results of fire tests on three small scale (1/4) steel frames. All three frames were two bay (each 1620 mm span) by one-storey (1400 mm high). One frame was unprotected, the second had fire protection on one of the two beams and the third had fire protection on both beams. Each frame was rigidly connected to the test laboratory and had welded beam to column connections. In all cases, the applied load was about half the resistance of the frame at ambient temperature. Fire exposure was by a gas fired furnace wrapped around each structural member. Measurements include temperatures at various locations in the steel frame and vertical and horizontal movements at the column heads and beam centres.

3.9.4   Model steel frame tests, Japan

In Japan, three series of tests were carried out to investigate the response of 2D and 3D model steel frames at elevated temperatures. Koike *et al.* (1982) reported the results of tests on the 2D steel frames. In these tests, temperatures in the steelwork were low enough that there was no damage to the test frame and the response of the structure was essentially linear and elastic. Ooyanagi *et al.* (1983) reported the results of tests on 3D steel frames, where only the beams were heated. Temperatures in the steelwork were high and caused buckling of some steel members. Hirota *et al.* (1984) reported a further series of tests on 3D model steel frames where the steel temperatures were high enough to induce structural damage.

In all three series of tests, the test structure was heated by an electrical furnace. Each test frame was installed with numerous strain gauges on the cold structural members and displacement transducers to enable accurate determination of changes in the axial load and bending moments of each steel member.

In the low temperature tests (Koike *et al.* 1982), significant increases in column bending moments were recorded when the adjacent beam was heated, due to the development of compressive forces in the restrained beam. In the tests reported by Ooyanagi *et al.* (1983) where the steel beams were heated to high temperatures, buckling of the beams was observed due to a combination of the thermally induced high axial load and bending moments in the beam. When bracing was used in the frame, the temperature at which the beam buckled was even lower due to the additional restraint of the beam's thermal expansion. After buckling, the axial load in the beam reduced due to reduced axial stiffness and length of the beam. Depending on the relative stiffness of the heated beam to the adjacent column and the heating condition, the adjacent cold column may also experience buckling due to the additional bending moment in the column generated as a result of the thermally induced axial load in the heated beam.

Results reported by Hirota *et al.* (1984) are similar to those by Ooyanagi *et al.* (1983). In these tests, they also reported buckling of some heated columns due to restraint by the adjacent structure.

Although buckling of the heated beams were observed in some of the above tests, there was no test information to further evaluate whether after shedding the axial load, the post-buckling response of the beam would have been able to sustain the fire attack for a much longer period. Without this information, it would not be possible to conclude whether the additional axial compression should be included in beam design.

The columns used in the above tests were usually short (1/4 scale) and the beams were reasonably long (1/2 scale). The axial restraint stiffness to a beam by the adjacent columns is from the columns bending stiffness and is related to a cubic function of the column length. The beam axial stiffness is linearly related to its length. Therefore, the axial restraint to a steel beam in a more realistic building could be about $4 \times 4 \times 4/2 = 32$ times smaller than the model restraint stiffness. Using this more realistic restraint stiffness, the restraint force in the steel beam could be much less. On the other hand, the axial restraint stiffness to a column in a realistic column may be much higher than from the model frames, thus more frequent column buckling may be encountered in realistic structures. The above tests do not provide sufficient information to assess the column post-buckling behaviour.

3.9.5   Tests of Kimura *et al.* on composite column assembly, Japan

To determine the failure condition of a structural member in fire, both the boundary and loading conditions of the member should be precisely defined. Due to difficulty and expenses of measuring strains at high temperatures, most of the previously described fire tests on frame assemblies do not provide adequate information to enable evaluation of the variation of forces in the heated members in frame tests. Among reports that do provide some such information, the results of Kimura *et al.* (1990) are particularly interesting.

In this report, a series of seven tests were carried out on beam–column assemblies in fire. Concrete filled square steel tubes were used as columns and their behaviour was the main interest of this study. Figure 3.17 illustrates the test arrangement, where rigid beam–column connections were used and the test assembly was fixed to the test laboratory.

Test parameters included the level of axial load and bending moment in the column and the amount of reinforcement. In each test, extensive measurements were taken of temperatures, displacements and strains at various

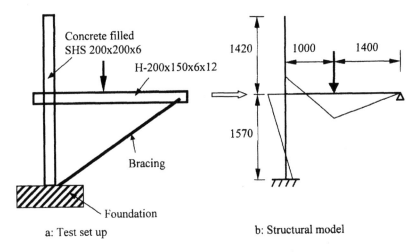

Concrete filled
SHS 200x200x6

H-200x150x6x12

Bracing

Foundation

a: Test set up

1420 1000 1400

1570

b: Structural model

Figure 3.17 Test set up and dimensions of Kimura *et al.* (1990).

locations of the test structure. The information from the strain gauge readings were particularly important as they enabled determination of variations of the bending moment and axial force in the test column.

Even though all test columns had initial bending moments, the observed axial deformation behaviour of each column was similar to that of an isolated column under pure axial loading. This may be explained by variations in the bending moment of the column during fire exposure.

Test observations indicate that due to weakening of the column, especially after local buckling of the steel tube, the column rigidity was low and as a result, the bending moment transferred from the beam to the test column was reduced substantially such that when approaching failure, the bending moment in the column was much less than 10% of the initial value. Significant reduction in the column bending moment appeared to coincide with the onset of local buckling in the steel tube. Figure 3.18 shows variations of the recorded bending moments in all test columns and the insert values give the times of local buckling in the steel tubes. Because bending moments in the column were low during a large part of the fire test, the column could be regarded as being axially loaded only.

The authors compared the influences of different test parameters on column fire resistance time and noticed that the column fire resistance was predominately affected by the initial axial load only, with very little effect from the initial bending moment and the amount of reinforcement. Thus, the influence of the initial bending moment in the column may be ignored.

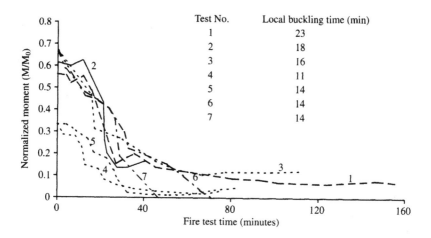

Figure 3.18 Variations of bending moment in restrained composite columns (from Kimura et al. 1990).

3.9.6 Parametrical fire testing of full scale structural assemblies, University of Manchester, UK

The large-scale fire tests on the eight-storey steel-framed composite structure in the BRE's Cardington laboratory (to be described in more detail in the next section) revealed many modes of structural behaviour in a complete building that have not been observed in individual or skeletal frame tests. The results of these tests identified the need for more targeted fire tests on steel structures. Two such test programmes are being carried out at the University of Manchester. One on restrained steel beams and one on restrained steel and composite columns.

3.9.6.1 Restrained beams tests (Fahad et al. 2000; Allam et al. 1999)

The objective of the restrained beam tests was to study the behaviour of steel beams (178 × 102 × 19 UB) connected to two support columns (152 × 152 × 30 UC) using different connections. A sketch of the test assembly is shown in Figure 3.19.

The beam was unprotected except at the top flange which was wrapped with ceramic felt to simulate the heat sink effect of concrete slabs. The columns were heavily protected so that they could be re-used. Test parameters included the type of connections (double web cleats and flush end-plates) and the level of loading. Temperatures and deflections were measured at various locations of the test assembly. No strain gauge was

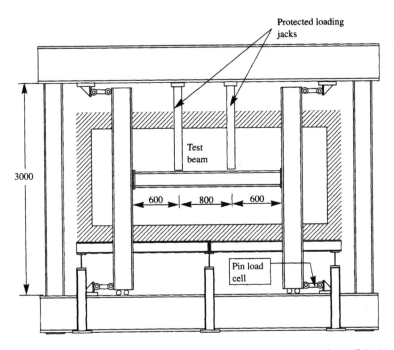

Figure 3.19 Schematic arrangement for restrained beam tests (from Fahad *et al.* 2000).

used, however, the forces in different members may be determined from horizontal reactions measured by the four pin load cells.

Initial investigations focused on the effect of using different types of connections on the failure temperature of the connected beam at different load levels. It was found that web cleat connections had very little influence on the behaviour of the beam until the beam came into contact with the columns. Flush end-plate connections were able to transfer a much higher bending moment to the column and the failure temperature of the beam was some 70 °C higher than a simply supported beam.

Catenary action was observed in some tests at very large beam deflections when tensile forces developed in the restrained beam (Fahad *et al.* 2001). However, no beam collapse was observed when in catenary action because the beam tests were terminated due to the difficulty of accommodating large deflections in the test furnace.

3.9.6.2 *Restrained columns tests (Hu et al. 2001)*

The previously mentioned tests of Kimura *et al.* (1990) can have significant impact on the fire resistant design of columns. If the initial bending moment in a column has to be included in design, as is normally required, design calculations will be complex and the column fire resistance will be low. On the other hand, if the observations of Kimura *et al.* are generally applicable, the initial bending moment in the column does not have to be considered in design.

The primary objectives of the restrained column tests were to investigate variations in bending moments of steel and concrete filled composite columns restrained by steel beams and the effects of changing the initial bending moments on failure temperatures of the restrained columns. The test set-up is shown in Figure 3.20. In this arrangement, a pair of short beams (about 1.5 m) are connected to the test column (about 3 m long) to induce bending about the column's minor axis by beam-column connections. The test beams are connected to a strong reaction frame using line roller supports to provide rotational restraint but to allow horizontal movement. The column ends are connected to the reaction frame using rollers so that the column is simply supported about the minor axis at both ends. Loads are applied to the column head and the two beam arms using hydraulic jacks.

Test parameters include three types of columns (UC 254 × 254 × 43, one concrete filled RHS 200 × 100 × 5 and one concrete filled RHS 200 × 100 × 12.5), two generic types of connection (flexible and rigid), different load levels and bending moments. The load level is determined by using the total axial load in the column. Within each load level, the applied loads on the two arm beams are varied so as to create different levels of initial bending moment in the column.

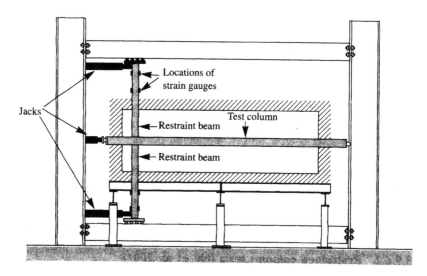

Figure 3.20 Schematic arrangement for restrained column tests (from Hu *et al.* 2001).

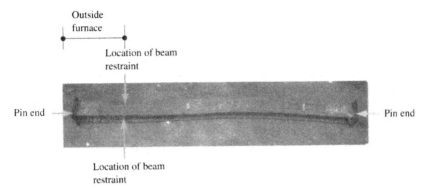

Figure 3.21 Deformation pattern of a restrained column (from Hu *et al.* 2001).

A gas-fired furnace is used to heat the structural assembly. Although the furnace can be controlled to increase the gas temperature according to any temperature–time curve, the heating rate is controlled at 20 °C/min so as to give sufficient test time (about 45 min) to record results for the unprotected structural assembly using UC columns. To maintain this initial load level, each hydraulic jack is connected to a relief valve through which the excess pressure in the hydraulic jack is released during the column expansion phase. When the column starts to contract, the hydraulic jacks are pumped to maintain the initially applied load. The test terminates when the column contracts so rapidly that the pressure in the hydraulic pumps cannot keep pace.

Extensive measurements are taken during each test to enable comprehensive determination of temperatures, forces and deflections in various parts of the structure assembly. Strain gauges are installed on the reaction frame to check and to ensure that the rollers between the arm beams and the reaction frame are functioning as expected and there is minimal friction force. Strain gauges outside the furnace are also installed on the arm beams. This enables determination of bending moments in the column.

At the time of writing this book, the test programme has just started. Nevertheless, from the appearance of a failed column (Figure 3.21), it is possible to detect that due to the restraint effect of the portion of cold column outside the furnace, the buckling length of the test column was less than the full length of the column. Detailed results of these tests and their implications on fire safety design will be published in due course.

3.9.7 Summary of fire tests on skeletal frames

A number of fire tests on skeletal steel frames and assemblies have been conducted by different researchers. The following observations may be drawn.

- Structural interactions between different members can alter the loading condition of any member. The effects of restrained thermal expansion and variations in stiffness should be considered.
- Local buckling appears to be a transitional phenomenon which does not affect the strength of a member in fire.
- Only a few tests can be used to provide detailed information on variations of forces in a structural member during the course of fire exposure.

3.10   FIRE TESTS ON COMPLETE BUILDINGS

Fire tests on skeletal frames of steel and composite structural members are necessary to understand interactions between different structural members and to appreciate differences between the behaviour of individual members and complete structural systems. However, in addition to the primary skeletal members, a complete structure also includes floor slabs, walls and other so-called "non-structural" members. A proper understanding of the structural behaviour of a complete building in fire can only be obtained if all such components are included. In other words, although extremely expensive, fire tests on complete buildings are necessary.

A number of fire tests on complete buildings have been carried out around the world. However, in terms of understanding the structural behaviour of a complete building in fire, the Broadgate fire accident (SCIF 1991) and the much publicized Cardington fire tests provide the most useful information. Other tests include the William street fire tests by BHP (Thomas *et al.* 1992) and the Collin street fire tests by BHP (Proe and Bennetts 1994) in Australia, the Basingstoke fire accident, the Churchill plaza fire accident in the United Kingdom and the fire tests (Anon 1986) in Germany. In these cases, steel temperatures were relatively low and the structural behaviour of these buildings can be adequately explained by the various modes of bending behaviour at small deflections already discussed with regard to tests on isolated elements.

3.10.1   Behaviour of the Broadgate building in a fire accident

Although not exactly a deliberate fire test, the experience of the Broadgate building in an accidental fire offered some insight into the likely behaviour of an unprotected, complete steel-framed composite structure in fire. The behaviour and the results of subsequent analysis of this fire (SCIF 1991) played a major role in the decision to carry out instrumented fire tests on the large-scale eight-storey steel-framed building in the UK BRE's Cardington laboratory (Section 3.10.2).

Figure 3.22 Damaged steelwork after the Broadgate fire (Reproduced from SCIF (1991) with the permission of The Steel Construction Institute, Ascot, Berkshire).

This major fire accident occurred in a 14-storey building under construction in Broadgate, London in 1990. The severe fire, which lasted $4\frac{1}{2}$ h, occurred when the building contractor's accommodation on the first floor level, which had been erected around the steel columns at that level, caught fire. The fire temperature reached 1000 °C during a substantial period of burning. The columns of the building, which passed through the contractor's accommodation at the heart of the fire, had not been fire protected. Neither were the floor beams/trusses protected. As a result, very high temperatures were attained in the steel work. The building contractor's accommodation was completely destroyed, but the steel frame survived the fire without a collapse. During the fire, the heavier columns survived undamaged and the lighter columns deformed in the heat and contracted by as much as 100 mm, see Figure 3.22. This behaviour of the steel frame could be attributed to the continuity of the structure; as soon as plastic deformation occurred in the failed columns (local failure), their loads were redistributed to other cooler members and the structure as a whole survived without a collapse. The structure was repaired in 30 days at a cost less than 4% of the total repair cost and no lives were lost.

3.10.2 Fire tests on an eight-storey steel-framed building in Cardington, UK

Without measured information, the experience of the Broadgate accident could not be directly used in fire safety design of steel-framed buildings.

Nevertheless, this accident became the catalyst in the decision to carry out full-scale structural fire tests in complete steel-framed buildings.

Opportunity to carry out large scale structural fire testing occurred when a former airship hanger was transferred to the BRE in 1989 and a strong floor was constructed for the purpose of large-scale testing on buildings (Armer and Moore 1994). This airship hanger is located at Cardington in Bedfordshire, UK and measures approximately 200 m long, 100 m wide and 80 m high. The strong floor measures 70 m by 55 m by 1.25 m deep.

Funding was obtained from ECSC (European Coal and Steel Community) and the UK's Department of Environment to carry out large-scale structural fire tests on complete buildings. An eight-storey steel-framed building was constructed within the Cardington airship hanger. Design was according to the UK's main steel design standard BS 5950 Part 1 (BSI 1990a) and checked for compliance with the provisions of Eurocodes (CEN 1992a,b). Design was carried out for a realistic office building in the Bedfordshire area where the Cardington laboratory is located. To make this test building realistic, the structural design was carried out by professional consulting engineers Peter Brett Associates and was free from input (or interference) by researchers.

Figure 3.23 Cardington steel framed building (Published courtesy of BRE).

Figure 3.23 shows this eight-storey steel-framed test building. The test building is a steel framed composite construction, using *in situ* concrete slabs supported by steel decking and in composite action with the supporting steel beams. Storey height is 4.285 m and there are five bays (5@9 m = 45 m) and three bays (6 + 9 + 6 = 21 m) on plan. The structure was designed as non-sway with a central lift-shaft and two-end staircases providing the necessary resistance to wind loads. The main steel frame was designed for gravity loads and the connections, which consist of flexible end plates for beam–column connections and fin plates for beam–beam connections, were designed to transmit vertical shear only.

The building was designed for a dead load of 3.65 kN/m² (including weights of composite slab, steel sections, raised floor, services and ceiling) and an imposed load of 3.5 kN/m³. However, at the time of structural fire testing, there were no raised floor, services and ceiling and the actual dead load was about 2.85 kN/m² (including weights of the composite slab and steel sections). Imposed loads were simulated using sandbags. Due to conservatism in the design load specifications, only about 2/3 of the specified imposed load was applied during the fire tests. Typically, 12 sandbags each of 1.1 ton were applied over an area of 9 m by 6 m, giving an uniform loading of 2.4 kN/m³.

The floor construction is of steel deck and light-weight *in situ* concrete composite floor, incorporating an anti-crack mesh of 142 mm²/m (T6@200 mm) in both directions. The floor slab has an overall depth of 130 mm and the steel decking has a trough depth of 60 mm. As a consequence of mistakenly placing the reinforcement mesh directly on top of the steel decking (see Figure 3.24), the anti-crack reinforcement device was not effective and cracks appeared along all the primary steel beams.

To rationalize sizes and to standardize connection details so as to reduce fabrication and erection costs, the entire structure used only three beam sections (356 × 171 × 51 UB as edge beams and the 6-m primary beams, 305 × 165 × 40 UB as interior secondary beams, 610 × 229 × 101 UB as the

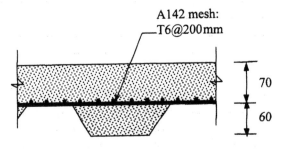

Figure 3.24 Position of anti-crack reinforcement in Cardington building.

9-m main beams and three columns (305 × 198 UC, 305 × 137 UC and 254 × 98 UC).

Fire test programme   Although the test building was used for a variety of tests, including structural fire tests, serviceability tests under static loading, floor collapse tests, dynamic loading tests, explosion tests and smoke movement tests, the structural fire tests were the main tests. The structural fire tests were carried out by Corus and BRE. In total, eight fire tests were carried out, including six compartment fire tests and two series of column tests. The two series of column tests were carried out to evaluate the effect of axial restraint on the behaviour of columns (see also Section 3.4.1). Tests were conducted by wrapping barrel furnaces around the columns. Results from these tests indicate that the restraint offered by the floor construction to the columns were low. The following descriptions will thus focus on the compartment fire tests. The Cardington fire tests have been widely described and more information may be found in Kirby (1997), Martin and Moore (1997); Newman *et al.* (2000); Wang (1998b, 2000b) and Wang and Kodur (2000).

3.10.2.1   Brief description of the compartment fire tests at Cardington

Locations

Figure 3.25 shows the locations and designations of the six compartment fire tests on plan. Corus carried out the restrained beam test, the plane frame test, corner compartment test 1 and the demonstration natural fire test. The BRE carried out corner compartment test 2 and the large compartment test.

Restrained beam test

This test involved heating a large part of a single secondary beam (305 × 165 × 40 UB) and a portion of the floor slab representing the concrete flange to the composite beam on the 7th floor. The beam span was 9 m and an 8 m long by 3 m wide gas-fired furnace was constructed beneath the beam. The steel beam was unprotected. The objective of this test was to provide some insight into the structural behaviour of a beam forming part of a complete building and to compare it with that from the standard fire resistance test on an isolated beam.

Plane frame test

In this test, a narrow strip across the entire width (21 m) supporting the fourth floor was heated by a gas-fired furnace built underneath the fourth floor, enclosing the three primary beams, the two internal and two external columns. The width of the furnace was 2.5 m. All beams were unprotected and all

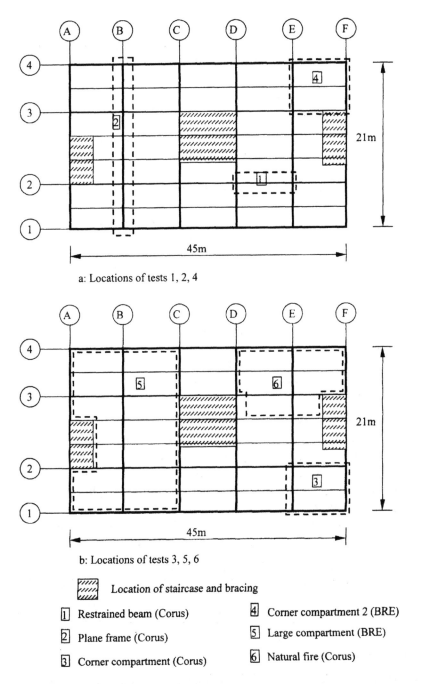

a: Locations of tests 1, 2, 4

b: Locations of tests 3, 5, 6

Location of staircase and bracing

1 Restrained beam (Corus)

2 Plane frame (Corus)

3 Corner compartment (Corus)

4 Corner compartment 2 (BRE)

5 Large compartment (BRE)

6 Natural fire (Corus)

Figure 3.25 Locations of Cardington fire tests.

columns were protected from the floor to the lowest beam flange. All connections were unprotected. The primary objective of this test was to provide experimental data to check capabilities of various computer programs which at that time were developed to analyse the behaviour of 2D steel structures in fire.

Corner compartment test 1

The first "real" compartment test was carried out by Corus. In this test, the floor structure of a corner block (9 m × 6 m on plan) on the second floor was enclosed in a fire compartment of approximately 10 m by 7 m on plan. Timber cribs were used to provide heating to simulate a natural fire. Fire load was 45 kg wood/m^2 floor area. Ventilation was provided by a 7 m wide opening with a moveable shutter in the elevation of the compartment to allow control over the burning rate and the compartment temperature. Perimeter beams of the building (including connections) and columns were protected. The internal primary beams and the secondary beam and their connections were exposed to the fire attack.

The objectives of this test were twofold:

1   to investigate the performance of a composite floor slab and interactions between different steel structural members; and
2   to provide some information for the validation of the Eurocode fire model in Eurocode 1 Part 1.2 "Actions on fire" (CEN 2000a).

Corner compartment test 2

This was the first of two tests carried out by BRE. The scope of this test was similar to that of corner compartment test 1 carried out by Corus. This test involved heating a corner block of 9 m by 6 m on the sixth floor so that the underside of the seventh floor slab was heated. A lightweight brickwork wall (gridline F) and two fire resisting partitions (gridlines 3 and E) formed the boundary of the fire compartment. The remaining boundary wall (gridline 4) was constructed of a double glazed window of about 2.5 m high running the full length of 9 m and sitting on a 1.5 m high brick wall. All columns were heavily protected and all beams were exposed to fire attack.

Fire load was provided with 40 kg of wood/m^2 floor area. The double glazing window was sealed and there was no additional ventilation. After ignition, the door was closed but there was no attempt to artificially seal the compartment.

Large compartment test

Being perhaps the largest fire test in the world, this test was carried out by BRE. It involved heating the entire width (21 m) and two of the five bays of the building (18 m) on the third floor. The fire compartment was bounded with a lightweight brickwork wall (gridline A) at one end and fire resisting partitions at the other (gridline C). Each two-bay side (gridlines 1 and 4) was bounded with two 6 m wide single glazed assemblies with a 6 m wide opening separating them. The opening was intended to simulate open windows in normal use and to avoid the need to break the windows during the fire test which had occurred in corner compartment fire test 2.

All columns were heavily protected, but all beams were exposed to the fire attack. The steel beams on gridline C were just outside the fire enclosure and remained cool during the fire test. Fire loading was again provided with 40 kg of wood/m^2 floor area. However, unlike in corner compartment fire test 2, the wooden cribs were arranged in such a way that each pile had a rather high weight but was separated by a long distance from other piles.

The objectives of this test were similar to the two corner compartment tests, however, the fire test was over a much larger compartment.

Demonstration test

The fire compartment in this test enclosed approximately 180 m^2, half of that in the large compartment fire test. The fire compartment was constructed on the first floor so that the second floor slabs were heated. All floor beams, including the edge beams were exposed to fire attack. Columns and connections were protected with ceramic fibre blanket material.

This was truly a real fire test and used real furniture (desks, chairs, filing cabinets, computer terminals, etc.) as combustible materials to provide an equivalent wood fire loading, in terms of calorific value of around 45 kg/m^2 floor area. A small number of wood cribs were included to aid fire development. Initial ventilation was provided by means of a "hit and miss" pattern of glazing to the building wall.

The objective of this test was to create a real fire scenario, to demonstrate many of the important lessons from the previous fire tests with regard to fire protection (or the lack of it) for steel framed structures.

3.10.2.2 *Observations from the Cardington fire tests*

Many lessons have been learnt from the Cardington fire tests and many more will be forthcoming after intensive analyses of the test data, supplemented by numerical modelling and more targeted experimental studies that are still on going. However, a number of apparently important observations are already having impacts on fire safety design of steel-framed structures and they will be presented here.

Fire behaviour

Of the four "natural" fire tests (corner compartment tests 1 and 2, large compartment fire test and demonstration fire test), the corner compartment fire test carried out by BRE presented the most interesting fire behaviour. After a short period of fire growth, the fire started to die down due to lack of oxygen. The stand-by fire brigade was asked to break a pane of the double glazing window. The influx of fresh air supplied the fire with oxygen and the fire started to grow again. However, the ventilation was still not sufficient and the fire appeared to die down again. It was only after the stand-by fire brigade was asked to break a window, the second time that the fire had sufficient oxygen to grow, leading to the expected fire behaviour of flashover to be described in Chapter 7. Figure 3.26 shows the recorded temperatures attained in various locations of the fire compartment. The two local peaks correspond to the two times when the fire brigade were asked to break a window.

It would appear to be possible to limit fire growth to the pre-flashover stage by using sufficiently strong windows in an sealed fire enclosure. If this could be done, structural damage would be minimal.

The fire exposure condition in the large compartment fire test was perhaps not realistic. In this test, adequate ventilation was supplied by openings in the windows on both sides of the fire compartment. However, because of the long distances between the wooden crib piles, each pile seemed to be burning in isolation. Therefore, although the combustion temperature directly above the burning pile reached the flashover temperature of about 600 °C, burning was limited to the available fire load area. The rate of heat release was low and as shown in Figure 3.27, the maximum fire temperature did not exceed 700 °C. At this fire temperature, the unprotected steel would usually have sufficient load bearing capacity in flexural bending without relying on the behaviour of the complete structure. However, in realistic fire situations, it is likely that fire loads will be more uniformly distributed so that higher fire temperatures will be attained.

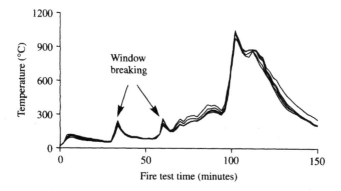

Figure 3.26 Recorded combustion gas temperatures in BRE corner test (Published courtesy of BRE).

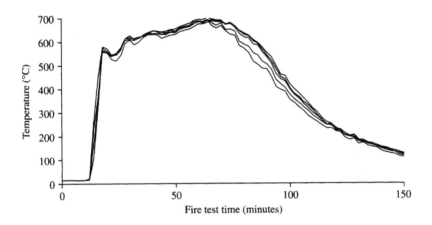

Figure 3.27 Recorded combustion gas temperatures in BRE large compartment fire test (Published courtesy of BRE).

Structural behaviour

The main purpose of these fire tests was to study the structural behaviour of a complete steel-framed building under realistic fire attacks so as to quantify the effects of structural interactions, load sharing and alternative load paths which cannot be assessed from fire tests on individual structural members. The Cardington fire tests have certainly achieved this objective and it is likely that outcomes from these fire tests will drastically change the fire safety design practice of steel-framed buildings in the future.

CONTRIBUTION OF NON-LOAD BEARING STRUCTURAL MEMBERS

In the Cardington test structure, each 9 m edge beam was linked to the beam on the floor above by two small angle posts at the two 1/3rd positions. The function of these "wind posts" was to secure the non-load-bearing walls and windows under wind loads. They were not designed to contribute to the vertical load carrying capacity of the beam that was designed as a simply supported 9-m span beam. However, the observed behaviour suggests that these non-structural members made a significant contribution to the survival of the edge beam during the fire tests. For example, Figure 3.28 shows the recorded maximum deflection–temperature relationship for the edge beam in corner compartment test 2. Very little vertical deflection was observed and this was attributed to the vertical support offered by these wind posts. In effect, the behaviour of the edge beam was close to that of a 3-span continuous beam instead of a 9-m simply supported beam as assumed in the design.

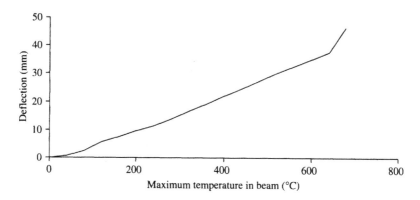

Figure 3.28 Recorded deflections of the edge beam in BRE corner test (Published courtesy of BRE).

LOCAL BUCKLING

In some tests (restrained beam, corner compartment test 1, demonstration test), local buckling of the lower flange and web of some fire-exposed beams was observed earlier in the fire test. This was clearly due to a combination of the large compressive force induced by axial restraint from the composite floor slabs and the hogging bending moments near the connections. However, local buckling did not seem to affect the global stability of these beams.

BEHAVIOUR OF BEAMS

Except in corner compartment fire test 1 where the perimeter beams (on gridlines 1 and F) were protected, beams in all other fire tests were unprotected. If any of these beams were treated in isolation as a composite beam, because of high temperatures attained in the unprotected steel beam, numerical simulations would have indicated that "runaway" deflection and beam failure would occur long before the maximum measured steel temperature. However, as indicated by Figure 3.29, which shows the deflection history of the fire exposed secondary beam in corner compartment test 2, there was very little sign of "runaway" deflection or structural failure at very high steel temperatures of about 900 °C. In fact, this figure shows an almost linear relationship, which seems to indicate that the floor system did not even lose much stiffness. Clearly, these fire exposed beams cannot be considered alone and should be treated as part of the floor structural system.

BEHAVIOUR OF COLUMNS

In the "plane frame" test carried out by Corus the columns enclosed in the fire compartment were protected, but only to the assumed position of the false

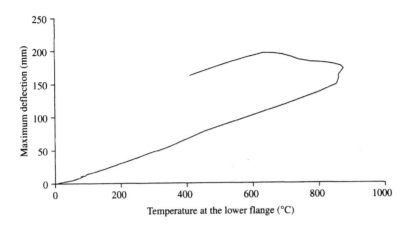

Figure 3.29 Recorded deflection of the secondary beam in BRE corner test (Published courtesy of BRE).

Figure 3.30 Fire protection of columns in Corus plane frame test (Courtesy of B. R. Kirby, Corus Fire Engineering).

Figure 3.31 Squashed column head after Corus plane frame test (Courtesy of B. R. Kirby, Corus Fire Engineering).

ceiling of about 200 mm below the lower flange of the larger central spine beam (610 UB), leaving a short length of column unprotected (see Figure 3.30). When the test was completed, it was observed that the short length of the unprotected columns was completely squashed. This is shown in Figure 3.31. As a result, the fire exposed 4th floor moved down by about 180 mm, taking with it all the floors above. This accidental behaviour has provided an important lesson: columns are critical members of a building structure and should be protected to limit their deflections. Even though an unprotected column may still have sufficient load bearing resistance to sustain the applied load in fire, excessive axial deformation of the column may lead to extensive damage to the entire structure, necessitating expensive repair cost. Fire protection to columns is relatively inexpensive and should be applied to include connections.

CONNECTIONS

Before the Cardington fire tests, some connection fire tests were carried out in isolation to obtain their moment–rotation relationships at elevated temperatures (cf. Section 3.8), with the intention of using the connection bending moment resistance to improve the fire resistance of the connected beams. However, connections in the Cardington frame behaved in a different way. In addition to bending action, the frame connection was subjected to a variable axial load due to a changing restraint acting on a varying thermal expansion of the connected beam.

During the heating stage, the thermal expansion of the beam was restrained and a compressive force was generated in the connection. During

Figure 3.32 Fractured connection after fire test (Courtesy of B. R. Kirby, Corus Fire Engineering).

the cooling stage, the composite floor slabs prevented contraction of the steel beam and caused tension in the connection. This tension caused fracture of the end plate and bolts in some tests (Figure 3.32), resulting in the loss of shear capacity in the beam. It is the sufficient load redistribution capacity of the floor slabs that ensured the stability of the floor structure. This type of connection behaviour was also observed after examining the Broadgate building (SCIF 1991).

BEHAVIOUR OF SLABS

Without any doubt, the most important lesson of the Cardington fire tests is the good performance of the floor slabs. In conventional analysis and design calculations, the common assumption is to treat floor slabs as the compressive flange of a composite beam in flexural bending. If this assumption were true, due to high temperatures attained in the supporting steel beams, the composite beams in the Cardington fire tests would have failed under the applied floor loads. However, the fact that none of these floor slabs showed any sign of imminent collapse suggests that the floor load was resisted by a different load carrying mechanism than flexural bending that was commonly assumed in practice. This load carrying mechanism has been identified as tensile membrane action (Wang 1996; 1997d) and will be discussed in more detail in Chapter 5.

COMPARTMENT (INTEGRITY) FAILURE

The Cardington test frame survived the various fire tests and retained its structural stability. However, it has to be appreciated that in fire resistant design, not only should a structure remain stable, it should also satisfy the integrity condition so that fire does not spread through openings in any structural member. Due to non-combustibility of steel and concrete, individual composite beams, columns and slabs cannot experience integrity failure. However, when these structural members experience large deflections, the connections between them may not be able to accommodate these large deflections and may cause integrity failure of the building.

In the large compartment test, the compartment wall underneath gridline C was designed to provide 2 h standard fire resistance and its head was detailed to accommodate a deflection of up to 50 mm. However, large deflections and rotations of the fire exposed beams caused buckling of these brittle partitions. In realistic situations, fire could have spread from the junction between the partition wall and underside of the floor slabs.

Figure 3.33 shows an opening in the floor slab around one column head in the demonstration fire test. This was caused by thermal contraction of the floor slab during the later stage of the fire test when cooling down. However, if these were to happen in a real building when the fire enclosure was still hot and producing smoke, there could be a danger of fire and smoke spread.

Figure 3.33 Floor opening around column head after Corus demonstration fire test (Courtesy of B. R. Kirby, Corus Fire Engineering).

It is clear that the large rotations and deflections experienced by floor slabs and beams will have implications on the design of the FR compartmentation system, the interface details between different components of a FR compartment, the positioning of fire resisting walls and the integrity of services etc. These aspects may influence the magnitude of deflections that can be tolerated at the fire limit state. This illustrates the need for a more rational fire safety engineering approach in which the whole building behaviour is taken into account.

REPARABILITY OF A FIRE DAMAGED BUILDING

Although reparability is not an explicit concern of FR design, it is expected that any fire damage should be commensurate with the extent of fire attack. This usually means that structural damage should be localized within the fire compartment area concerned. This was the case in all Cardington fire tests except for the plane frame test. In all other tests, repair of the fire damaged structure would have been relatively simple, involving cutting out the fire damaged floors and replacing the fire damaged steel sections.

In the case of the plane frame test, repairing the fire damaged structure would be more intensive and more disruptive in terms of building occupancy. As described previously, in this test, the squashing of a short unprotected length of the column head brought down all the floors above. Repair of the fire damaged structure would probably involve the difficult task of jacking up the two bays of the entire structure from the third floor to the eighth floor and replacing the fire damaged steelwork.

3.10.2.3 Design implications of the Cardington fire tests

The Cardington fire tests have clearly demonstrated the superior fire performance of complete structures that cannot be evaluated from standard fire resistance tests on simply supported individual elements. This was the primary objective of the Cardington fire tests and further research studies are being undertaken to translate the Cardington observations into practical design guides.

There is no doubt that the observed superior behaviour of composite floor slabs in fire represents the most important finding of the Cardington fire tests. A proper understanding of the true slab behaviour will enable extensive use of unprotected steelwork.

However, the Cardington fire tests were equally important in revealing some of the concerns caused by using unprotected steelwork when it interacts with other building components. Even though the Cardington building remained structurally stable, large deflections and rotations associated with using unprotected steelwork caused integrity failure which may lead to fire spread and extensive fire damage in realistic situations. Clearly, further studies are necessary to alleviate these problems so that full benefits of the Cardington fire tests can be utilized.

3.11 CONCLUDING REMARKS AND SOME SUGGESTIONS FOR FURTHER EXPERIMENTAL STUDIES

This chapter has presented a review of a large number of fire tests on steel structures and their main observations. These tests may be divided into two types: (1) tests on statically determinate, simply supported structural elements with well-defined loading conditions; and (2) tests on structural assemblies and complete structures.

Consider first individual elements. The modes of bending behaviour that are considered at ambient temperature can be directly applied under fire conditions, provided suitable allowances are made for degradations in mechanical properties of the constituent materials and thermal deformations.

With regard to tests on structural assemblies or complete structures, the essential effect of a fire is to make the loading and boundary condition of any member in a structure variable. Moreover, due to redundancy in the structure, it is very likely that some structural members can undergo very large deflections without causing a structural collapse. Further experimental studies are necessary to understand the behaviour of structural elements at large deflections and of interactions in structural assemblies.

Of course physical testing of structures is time consuming and expensive, this is especially true of structural testing under fire conditions. It is therefore not surprising that these shortcomings have often been used as justifications for the development of numerical models to replace physical testing. However, despite the proliferation of numerical models (a few well known ones will be reviewed in the next chapter), physical testing continues to be important in developing new understandings. It can be argued that without the Cardington fire tests, it would have been almost impossible to achieve today's understanding of whole structural behaviour in fire. However, the Cardington fire tests should not be seen as the end of physical testing. Instead, they have revealed the extent of our lack of knowledge in many areas and should encourage more physical testing. Of course, it would be impossible to acquire the financial support to conduct such large-scale fire tests again. However, a proper analysis of the structural behaviour should help in identifying more targeted tests on individual structural elements and structural assemblies involving only a few structural members. If further experimental studies are to be carried out, careful attention should be paid to support conditions of the test specimens to reflect realistic interactions between different structural members in complete structures.

The following list is by no means exhaustive, but it points out a few directions of further experimental studies under fire conditions.

- Local buckling of thin-walled steel structures: this area is generally not well researched, in particular, tests should concentrate on thin-walled members with non-uniform stress and temperature distributions.
- Axially restrained columns: the behaviour of an axially restrained steel column during expansion is now well understood. However, the behaviour of an axially restrained composite column can be different and requires further experimental studies. Particularly, future tests should include the post-buckling behaviour of restrained columns.
- Rotationally restrained columns: the bending moment transfer in rotationally restrained columns is being studied. The problem of effective length of rotationally restrained columns needs further studies. A particular problem is concerned with the effective length of concrete filled composite columns. The well-known conclusions for steel columns (cf. Section 5.3.4) may not necessarily apply to concrete filled composite columns due to uncertainty over the location of local buckling in the steel tube.
- Lateral torsional buckling in unrestrained beams: considering that LTB is one of the main topics in the design of steel beams at ambient temperature, it is surprising that LTB is a very poorly researched topic in fire. It is rare for LTB to govern the behaviour of a steel beam in fire, however, even as a transitory phenomenon, it is important to understand how LTB may influence the ultimate strength of a steel beam.
- Axially restrained beams: if a steel beam is axially restrained, it can be expected that catenary action will govern its ultimate collapse. However, on the way to ultimate collapse, the beam may undergo a series of other transitory failures, including local buckling, LTB and a plastic hinge mechanism. These modes of behaviour may influence the deflection and catenary forces in the beam at collapse, which in turn determine the anchor resistance that should be provided by the adjacent structure. Some computer programs may simulate this range of behaviour, however, experimental results are required to establish the beam collapse criteria. It is quite possible that the utilization of catenary action in steel beams may lead to the ultimate goal of the steel industry by completely eliminating fire protection to all steel beams.
- Columns under combined axial load and bending moments: if catenary action develops in a beam that is connected to a column, the column should have sufficient strength to resist the catenary force in the beam. This tensile force can produce large bending moments in the column. Previous experimental studies on columns have largely focused on

axially loaded ones. The behaviour of a column with combined axial load and bending moments has never been rigorously evaluated.

- Connections: studies of connections should progress on two fronts. The recently started tests on isolated connections should continue, however, they should be broadened to include other types of connections, e.g. connections to tubular columns, and other design parameters, e.g. different fire exposures. These tests are necessary to establish some basic understanding of the connection behaviour in fire and to provide experimental information for numerical modelling. Ultimately, connection testing in fire should include the effects of variable axial loads. These tests may necessitate the use of structural assemblies.

- Slabs: the strength of a slab in tensile membrane action is making immediate impact on fire safety design of steel-framed structures. However, the design guidance has been derived from slab tests that did not experience collapse and that were conducted at ambient temperature. For such an important design check, it is essential that collapse tests be carried out in fire so that the ultimate strength of the slab can be precisely calculated.

Chapter 4

Numerical modelling

Fire tests on structures are expensive and time consuming. Because of this, the development of accurate predictive methods to simulate the behaviour of steel structures in fire has long been regarded as desirable. Earlier efforts concentrated on predicting the fire resistance time of isolated members to mimic directly the standard fire resistance test. Although some of these earlier attempts could also be used to give detailed information for the performance of a structural member in fire, e.g. variations of stresses and displacements as functions of the standard fire exposure time, the majority of these studies were only interested in predicting the ultimate load carrying capacity of the member at a certain fire exposure time or the standard fire resistance time of the member under the initially applied loading condition. These studies adequately served their purpose of quantifying the ultimate limit state of isolated structural members under fire conditions, however, they cannot consider the performance of complete structures. Moreover, these programs have limited capabilities and cannot simulate any advanced structural effects other than the basic flexural bending behaviour at small deflections.

For a complete structure, the performance of a structural member in fire is dependent on its interactions with other structural members. Therefore, to study the fire performance of a structural member as part of a building structure, it is necessary to consider the structure as a whole. Indeed, recent numerical developments are mainly concerned with the analysis of complete structures. To adequately model a structure that can have general layout, loading, boundary and fire exposure conditions, the versatility of the finite element approach has made it the preferred method of most researchers.

This chapter provides a brief review of some finite element analysis programs that are used in the field of steel structures in fire. It is not the purpose of this chapter to discuss methodologies and implementations of finite element procedures. Interested readers should consult many excellent and authoritative textbooks on this subject such as Bathe (1996) and Zienkiewicz and Taylor (1991). Instead, this chapter will provide an assessment of a few currently better known computer programs for the analysis of steel framed structures in fire. This chapter will start with the requirements of a computer program to adequately quantify the various phenomena of structural behaviour observed from the fire tests described in the previous chapter. A brief outline is then given of each program commenting on its capabilities, applications, advantages and disadvantages. The programs included in this assessment are ADAPTIC (Izzuddin 1991, 1996; Song *et al.* 1995, 2000; Song 1998), Finite Element Analysis of Structures at Temperatures (FEAST) (Liu 1988, 1994, 1996), SAFIR (Franssen *et al.* 2000), VULCAN (Bailey 1995, 1998a; Huang *et al.* 1999a, 2000a,b; Najjar 1994; Najjar and Burgess 1996; Rose 1999; Saab 1990; Saab and Nethercot 1991), ABAQUS and DIANA. The first four are specialist programs dedicated to analysing steel structural behaviour in fire and the last two are commercially available general finite element packages that have been successfully applied to structural analysis in fire.

4.1   REQUIREMENTS OF A COMPUTER PROGRAM

Numerical modelling of the behaviour of a steel structure under fire conditions has undergone enormous changes in recent years. Initially, these models were developed to provide an inexpensive alternative to the standard fire resistance test. Nowadays, they are becoming more widely used as an indispensable tool for researchers to gain a deep understanding of various complex modes of structural behaviour and their interactions and to help designers to select appropriate strategies in the fire resistant design of steel structures.

To ensure confidence in the suitability of a computer program, it should be thoroughly checked against experimental results. As indicated in Chapter 3, until very recently, experimental studies of structural behaviour in fire have concentrated on limited aspects of isolated structural elements. Therefore, confirmation of a computer program's ability to model these aspects does not necessarily extend its validity to other modes of structural behaviour and structural interactions. From experimental observations described in the previous chapter, especially from the Cardington fire tests, one can put forward a list of requirements that a computer program should meet. Of course, not all computer programs will be able to meet all these requirements. Therefore, applications of a computer program should be limited to its validated range.

The effects of fire attack on a structure are two-fold: (1) to degrade mechanical properties of the constituent materials (steel and concrete); and (2) to induce interactions wherever restraints are present. Within a complete structure, the loss of load carrying capacity of a structural member in fire often necessitates redistribution of loads to the adjacent structure. Therefore, numerical modelling of the behaviour of a complete structure in fire should be tackled on two fronts: (1) addressing the local behaviour and load carrying capacity of isolated elements; and (2) on a global level, interactions and redistributions of load in complete structures.

4.2 MODELLING STRUCTURAL BEHAVIOUR IN FIRE ON AN ELEMENT LEVEL

4.2.1 Beams and columns

From observations in Chapter 3, the following modes of structural behaviour should be included in modelling the behaviour of beams and columns in fire:

- local buckling;
- flexural bending and shear;
- lateral torsional buckling;
- distortional buckling; and
- catenary action at large deflections.

The modelling of beams and columns are usually performed using one-dimensional line elements. With appropriate selection of the degrees of freedom for a line element, in plane bending, shear and LTB can be modelled. To simulate catenary action, the program should be able to follow the geometrical non-linear behaviour of the element to very large deflections. Local buckling and distortional buckling can occur in fires, especially in thin-walled plates. One-dimensional line elements cannot simulate local buckling and distortional buckling; detailed shell or brick elements should be used.

4.2.2 Connections

Connections refer to structural components that join different structural members as in beam to column connections or at interfaces of different materials such as shear connectors in a composite beam or bond in a composite column or slab.

At ambient temperature, a beam to column connection is usually represented by a spring element linking the respective beam and column nodes. Since the connection is under predominantly bending action, moment–rotation relationships are sufficient to represent the connection behaviour. Under fire conditions, the connection behaviour can no longer be adequately represented by moment–rotation relationships. Large variable axial forces can be induced in the connection under fire conditions and the connection is no longer under predominantly bending action. The development of a variable axial force is due to the restraint to thermal expansion during heating and to thermal contraction during cooling. This axial force, in combination with the bending moment and shear force in the connection, can cause fracture of some connection components (bolts, end plate etc.). Therefore, connection modelling should use more detailed elements such as shell, bolt and solid elements.

The connection between the steel section and the concrete flange in a composite beam is usually by shear connectors. Depending on the ductility and number of shear connectors and the heating and cooling regimes of the steel and concrete, shear connectors may initiate failure. The steel and concrete components may have to be treated separately with link elements to represent the shear connectors.

The behaviour of the steel–concrete connection in a composite column is complex. Depending on the bond strength, steel and concrete may displace relative to each other. Again link elements should be provided between the steel and the concrete components to simulate bond at the interface. However, due to a lack of information on the characteristics of bond at elevated temperatures, numerical simulations are likely to assume composite behaviour without considering the effect of bond.

Bond between the concrete slab and steel decking in a composite slab should also be considered so as to evaluate the contributions of the steel decking. However, this is unlikely to be attempted due to the complexity of having to simulate moisture movement and steam pressure inside the concrete.

4.2.3 Slab modelling

The strategy for modelling slab behaviour depends on the function of the slab. If it is merely to enhance the bending resistance of a steel beam, the slab may simply be treated as part of the beam element, for example as the compression flange of a composite beam. However, in a real building, slabs can distribute floor loads to the surrounding structure in two way bending. At large deflections, compressive and tensile membrane actions may be activated. The only realistic way of slab modelling is to use shell elements.

4.3 MODELLING STRUCTURAL BEHAVIOUR IN FIRE ON A GLOBAL LEVEL

The main difference between the behaviour of a complete building structure in fire and that of an isolated structural element in fire is that the complete building is highly redundant and can have many load paths. Load redistribution can, and will, almost certainly occur and alternative load paths will be activated when failure (whatever the definition) of some structural elements occur. Therefore, in addition to modelling large deflections on the element level as already mentioned above, robust solution strategy should be built into the computer program to track progressive failure.

It is not likely that a complete structure can be modelled using only one type of finite element, i.e. detailed shell elements only will be costly and not practical, but beam elements only may not model all modes of structural behaviour. Therefore, a computer program should include a large library of finite elements so that the most computationally efficient combination of finite elements is used according to the anticipated functions and modes of behaviour of different structural elements.

4.4   OTHER GENERAL MODELLING REQUIREMENTS

At high temperatures in fire, the stress–strain relationships of material become highly non-linear. It is essential that a computer program should include material non-linearity.

Temperature distributions in a structure under fire conditions will in general be non-uniform. The preparation of temperature distributions for structural analysis can be a tedious process. Ideally, structural analysis should be fully integrated with thermal analysis so that temperature results from the thermal analysis can be directly imported into the structural analysis. However, this can be difficult due to incompatibility between different finite elements for thermal and structural analyses and the fact that some elements which must be included in the thermal analysis (e.g. fire protection) may not be included in the structural analysis.

Modelling the behaviour of a structure under fire attack is a specialist task, requiring the user to have a thorough understanding of various complex structural interactions at high temperatures and experiences of operating non-linear finite element packages. To alleviate the problem of misuse, good technical support (including comprehensive documents) by the program developer is essential. In this regard, good pre- and post-processors should be provided to reduce the task of data preparation and to aid interpretation of simulation results. In particular, the users of such programs (usually structural engineers) are more familiar with interpreting forces (axial force, bending moment, shear) rather than stresses. Therefore, when shell or more detailed (i.e. brick) elements are used, facilities should be provided in the computer program to convert stresses into resultant forces.

4.5   A BRIEF REVIEW OF SOME EXISTING COMPUTER
PROGRAMS

All computer programs may be divided into two broad groups: (1) specialist programs that have specifically developed for steel structural analysis under fire conditions; and (2) generally available commercial programs that have been adapted for structural fire analysis. The former are usually developed at a university or research organization based on outcomes of a number of research projects. As such, these programs can be expected to have a history of validation study for analysing structural behaviour under fire conditions. Since they are specialist programs, their input and output will usually be brief and directly related to the fire problem.

A comprehensive review of the capabilities of fire dedicated thermal and structural analysis programs up to 1990 was provided by Terro *et al.* (1991). Both steel and concrete structural analysis programs were included in the assessment. The thermal analysis programs reviewed include FIRES-T3 (Iding *et al.* 1977a), TASEF (Wickstrom 1979), SUPER-TEMPCALC (IFSD 1986), STABA-F (Rudolph *et al.* 1986), CEFICOSS (Franssen 1987; Schleich *et al.* 1986), IMPERIAL COLLEGE (Terro 1991). The structural analysis programs include: FASBUS-II (Jeanes 1985), FIRES-RCII (Iding *et al.* 1977b), STABA-F (Rudolph *et al.* 1986), CONFIRE (Forsen 1982), STEELFIRE (Forsen 1983), CEFICOSS (Franssen 1987; Schleich *et al.* 1986), ISFED (Towler *et al.* 1989), BFIRE (Newman 1995), FIRESTRUCT (OAP 1985), IMPERIAL COLLEGE (Terro 1991), INSTAF (El-Zanaty *et al.* 1980; El-Zanaty and Murray 1980, 1983), ABAQUS, WANG (Wang 1992; Wang and Moore 1992, 1995) and SOSMEF (Jayarupalingam 1996; Jayarupalingam and Virdi 1992). Milke (1992) assessed the capabilities of the two thermal analysis programs TASEF and FIRES-T3. Franssen *et al.* (1994) compared predictions of five structural fire codes for steel elements. These five codes are CEFICOSS, DIANA, SAFIR (Franssen *et al.* 2000), LENAS-MT (Kaneka 1990) and SISMEF of CTICM, France.

Apart from ABAQUS and DIANA, all programs reviewed above were specialist fire dedicated programs, and most of these programs were still at a very early stage of their development. Since the review, many of the specialist programs ceased to be further developed. A few of the programs have been continuously updated by successive researchers and other new programs have also been developed. In this book, only a selection of programs will be reviewed. These programs are ADAPTIC, FEAST, SAFIR, VUL-CAN, ABAQUS and DIANA. Of these programs, ABAQUS and DIANA are commercially available general finite element packages that have been successfully adapted for fire related structural analyses. The other four programs are specialist ones. The main reason for selecting these programs in this review is that these programs are being actively developed and used in major centres of steel structural fire research in the United Kingdom and in Europe, and to some extent in other parts of the world.

4.5.1 ADAPTIC

The computer program ADAPTIC was initially developed by Izzuddin (1991) at the Imperial College, London to study the non-linear dynamic behaviour of framed structures at ambient temperature. This was extended to include fire and explosion effects on steel framed structures (Song *et al.* 1995, 2000; Izzuddin 1996; Elghazouli *et al.* 2000). Later developments by Song (1998) extended the capability of the program to deal with reinforced concrete floor slabs.

For frame analysis, this program uses two types of beam elements: quartic elements and cubic elements. In a quartic element (which is used in elastic analysis only), higher order quartic functions are used for the transverse displacements. This allows elastic buckling to be accurately analysed using only one quartic element per member. More cubic elements will be necessary to achieve the same accuracy as one quartic element. Therefore, although the quartic element formulations are computationally more demanding per element, computing effort is saved by using one element per structural member.

Under fire conditions, due to material non-linearity at elevated temperatures, more elements per structural member are required for accurate modelling. In this case, it is more advantageous to use elasto-plastic cubic elements. The program ADAPTIC has a very useful feature to automatically re-mesh the structure to allow more cubic elements to be introduced for the structural member concerned if conditions of the member are beyond the applicability of quartic element formulations (Song *et al.* 2000). Figure 4.1 shows this automatic re-meshing scheme. A typical ADAPTIC analysis involves the following procedure:

1 Start the analysis with one elastic-quartic element per structural member and carry out the analysis in an incremental way.
2 After achieving convergence, check whether any quartic element has exceeded its range of applicability at the pre-defined locations.

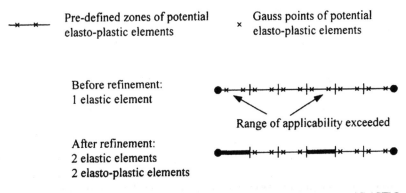

| —×— ×— | Pre-defined zones of potential elasto-plastic elements | × | Gauss points of potential elasto-plastic elements |

Before refinement:
1 elastic element

Range of applicability exceeded

After refinement:
2 elastic elements
2 elasto-plastic elements

Figure 4.1 Checking and refinement of a typical elastic element in ADAPTIC (Reprinted from the *Journal of Constructional Steel Research*, "An integrated adaptive environment for fire and explosion analysis of steel frames – Part I: analytical models". Vol. **53**, pp. 63–85, Song *et al.* (2000) with permission from Elsevier Science).

3   If the range of applicability of any zone is exceeded, re-mesh this zone using cubic elements. The remaining unaffected zones of the original element are still modelled using elastic-quartic elements.

For fire analysis, re-meshing is triggered by any of the following conditions:

- if mechanical strain exceeds the material yield strain at elevated temperatures;
- if thermal expansion is no longer linearly related to temperature; and
- if there is a significant variation in the elastic Young's modulus due to temperature variation.

Under fire conditions, due to reduced loading, it is rare that any of the unexposed structural members will be loaded beyond its elastic limit. Therefore, during the course of fire attack on a structure, it is likely that only a small number of structural members will be involved in non-linear behaviour.

The ADAPTIC program was originally developed for dynamic analysis of steel frames. A particularly useful feature of this program is its ability to simultaneously perform dynamic and fire analysis. Static analysis is normally sufficient for modelling the behaviour of a structure under fire conditions, however, using quasi-dynamic analysis, it is possible to deal with the situation where elevated temperatures lead to a temporary loss of structural stiffness but the structural stiffness is regained at large displacements. In other words, the program has a robust strategy to deal with progressive failure of a structure.

4.5.1.1   Applications of ADAPTIC

The capabilities of the ADAPTIC program to deal with framed structures have been extensively checked against results of fire tests and the program has been shown to produce accurate results (Izzuddin et al. 2000). At the time of writing this book, slab elements of the program are still being developed. To model the behaviour of a slab, ADAPTIC uses a grillage representation of the slab (Izzuddin and Elghazouli 1999; Elghazouli et al. 2000). This approach has been successfully used in modelling the Cardington fire tests. For example, Figure 4.2 shows a comparison between the Corus restrained beam test results (cf. Section 3.10.2.1) and simulations by ADAPTIC (Elghazouli et al. 2000).

A particular capability of the ADAPTIC program is to model the effect of fire after an explosion. For example, using ADAPTIC (Izzuddin et al. 2000), analysis was carried out to study the behaviour of a three-storey frame. It was shown that due to damage (plastic strain) accumulated in the structure during an explosion, the resistance of the structure to a subsequent fire is lower than that of the structure without the explosion. Without explosion, the frame could resist a high steel temperature of 894 °C. When explosion was considered, the failure temperature was reduced to a much lower value of 642 °C.

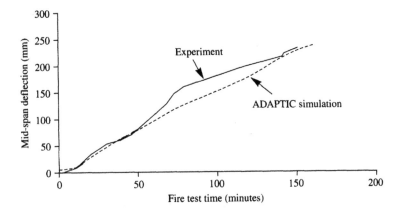

Figure 4.2 Comparison between ADAPTIC prediction and measured results for Corus restrained beam test (Reprinted from the *Journal of Constructional Steel Research*, "Numerical modelling of the structural fire behaviour of composite buildings". Vol. **35**, pp. 279–297, Elghazouli *et al.* (2000) with permission from Elsevier Science).

4.5.1.2 Limitations

The ADAPTIC program has a number of unique features which are not present in other specialist programs and these make it an exceptional program with a much wider range of potential applications. Nevertheless, at present, there are still a few aspects of the structural behaviour that have not been included in the ADAPTIC program. The ADAPTIC program has one-dimensional beam elements for modelling steel frames and shell elements for modelling concrete slabs. It cannot be used to deal with problems of local buckling and distorsional buckling. Also its one-dimensional beam element does not include the warping degree of freedom so that it cannot simulate lateral torsional buckling. Its beam elements are for steel structures only and cannot deal with composite construction. Detailed connection behaviour cannot be studied using ADAPTIC. As with other purely structural analysis programs, its temperature input is from other sources.

4.5.2 FEAST

Finite Element Analysis of Structures at Temperatures (FEAST) has been developed at the University of Manchester by Dr T. C. H. Liu. This program was originally developed to study the detailed structural behaviour of steel portal frames at ambient temperature, as part of Dr Liu's Ph.D. research (Liu 1988).

The computer program has recently been extended to analyse the behaviour of steel structures subject to fire attack (Liu 1994, 1996). The stress–strain relationships of steel at elevated temperatures are based on those obtained by Kirby and Preston (1988). For concrete, the elevated temperature constitutive relationships are according to Kasami (1975). Since concrete in a composite connection is mainly subject to tension and light compression, the concrete crack–crush model of William (1974) was adopted and modified to simplify the material behaviour in a low compression region.

The program's library of finite elements includes 8-noded shell elements, 8- or 14-noded solid elements, bolt, gap and contact elements. Therefore, all modes of local behaviour listed in Section 4.2 for steel beams and columns can be analysed. At present, only linear elastic beam elements are included in the program. Thus it is not practical to use this program to analyse the non-linear behaviour of large-scale steel framed structures.

A particular useful feature of this program is in modelling bolts (Liu 1994, 1999b). In a bolted connection at high temperatures, a bolt may become slack when it expands at increasing temperatures. Using conventional one-dimensional beam or bar element, since no slack is allowed, the adjacent steel plates would present an axial restraint to the thermal expansion of the bolt and would induce a spurious compressive load in the bolt. To resolve this problem, a three-noded bolt element has been developed.

The program allows the user to select any combination of load control, temperature control, displacement control with constant load and variable temperature, and displacement control with constant temperature and variable load. This feature is particularly useful for post-buckling analysis.

Finite Element Analysis of Structures at Temperatures (FEAST) employs the frontal-solver technique. Unlike the Gauss elimination method, the frontal-solver technique does not have to terminate when a structure encounters a local failure which causes the diagonal element of the stiffness matrix to become non-positive. Therefore, this program can be used to find solutions for post-failure analysis.

The FEAST program has a very useful restart feature. After the program has executed the required number of load/temperature/displacement increments, data are stored in a number of temporary files. If the user wishes to continue more steps, these additional steps can simply be appended to the original input data file. The program will open the temporary files and restart from the first new increment. This avoids having to go through the previous steps in a reanalysis of the problem and gives the user flexibility to change the control strategy and control steps.

The FEAST program has a functional window based pre-processor to help data preparation. Its post-processor is straightforward and the user can plot various stress and displacement distributions. At present, there is no facility to convert stresses in shell elements to resultant forces.

4.5.2.1 Applications of FEAST

This program has mainly been applied to studying the detailed behaviour of steel framed connections (Liu 1999b) and the influence of connections on

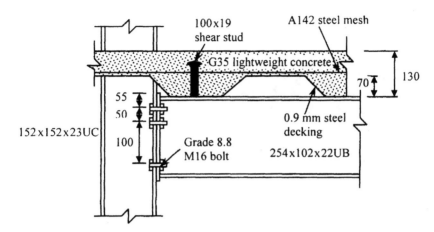

Figure 4.3 Details of a composite connection tested by Leston-Jones (1997).

steel beams under fire conditions (Liu 1996, 1998, 1999a). Recently, the program is being applied to studying detailed buckling problems in thin-walled steel structures, LTB in steel beams, shear buckling in deep and cellular beams and catenary action in steel beams under fire conditions.

The ability of the FEAST program to analyse the detailed behaviour of steel and composite connections at elevated temperatures has been confirmed by extensive comparisons against available results of connection tests at elevated temperatures (Lawson 1990b; Leston-Jones 1997; Al-Jabri *et al.* 1998). For example, Figure 4.3 shows details of a composite connection tested by Leston-Jones (1997). The FEAST finite element mesh of the connection is shown in Figure 4.4. Figure 4.5 compares the measured temperature–rotation relationships with simulation results. Agreement between these two sets of results is remarkable and the FEAST program can be regarded to be able to predict detailed connection behaviour at elevated temperatures very accurately.

4.5.2.2 Limitations

At present, the FEAST program is capable of accurately predicting the detailed behaviour of steel members and steel and composite connections under fire conditions. Various buckling phenomena in a steel member can be simulated. Since the beam element formulation is linear elastic, it is not practical to use this program to analyse the non-linear behaviour of large-scale steel frames with many members. Also, the concrete constitutive model is not robust and it is not suitable to simulate composite structural behaviour. The thermal analysis program FIRES-T3 (Iding *et al.* 1977a) has been

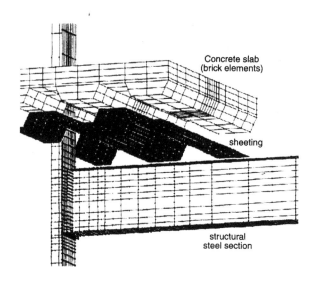

Figure 4.4 FEAST finite element mesh of a composite connection (from Liu 1999b).

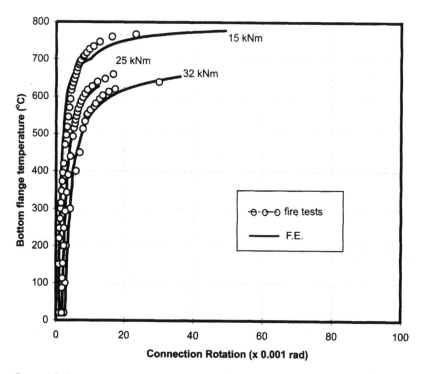

Figure 4.5 Comparison between connection test results and FEAST simulation (from Liu 1999b).

linked to FEAST, but more work is required to address the problem of finite element compatibility.

4.5.3 SAFIR

The program SAFIR (Franssen *et al.* 2000) has been developed at the University of Liege, Belgium by Franssen and later developments were supervised by Franssen. The predecessor to the SAFIR program is the well-known computer program CEFICOSS (Franssen 1987).

SAFIR can be used for both thermal analysis and structural analysis at elevated temperatures. At present, the thermal and structural analyses are not fully coupled. The user has to carry out thermal analysis first for each part of the structure and then prepare a library of temperature files to be used for subsequent structural analysis.

The SAFIR program has a finite element library of 2D solid elements with 3 or 4 nodes; 3D solid elements with 6 or 8 nodes; one-dimensional beam elements; 4-noded shell elements and truss elements. Various material constitutive laws for steel and concrete are included in the program and the user can also input their own material models. Thermal analysis is performed using triangular or rectangular 2D solid elements or 3D solid elements. Structural analysis is performed by using truss elements, beam elements, shell elements or 3D solid elements. At present, the 3D solid element can only simulate elastic behaviour.

The beam element uses the conventional 6 degrees of freedom at each node. In order to deal with non-uniform torsion and warping, the program can be used to carry out a torsional analysis to obtain the torsional stiffness of the cross-section. The torsional properties of the cross-section together with temperatures are used as input data in subsequent structural analysis.

The use of shell elements (Talamona and Franssen 2000) is a relatively new addition to the program. This allows local buckling to be simulated. At present shell elements are not linked to other types of element and have to be used on their own. In addition, the shell element has mainly been developed for analysing steel structures. Its ability to simulate concrete structures has not been validated.

The arc-length method (Crisfield 1991) is included in the program to analyse post-buckling behaviour. However, the arc-length method is implemented in such a way that at present, only simple structures can be analysed (Franssen 2000). In an analysis, initial loads are applied on a structure and kept constant. The structural temperature is then increased until first failure. At this stage, the temperature at first structural failure is kept constant and the arc-length method is applied in the conventional load-deflection domain to find an equilibrium position for the structure with reduced loads. From the new equilibrium position, the structural temperature is increased again.

The SAFIR program is supported with window based pre- and post-processors. For thermal analysis of a steel member or a composite member with a concrete slab on top of a steel section, the pre-processor has been specially adapted for very easy preparation of data. Unlike other specialist programs, the SAFIR program has been used by a number of organizations around the world and there is a small user group.

4.5.3.1   Applications of SAFIR

The SAFIR program has mainly been used to perform simulations for framed structures and for the calibration and development of simple design rules for beams and columns (Franssen *et al.* 1995; Vila Real and Franssen 2000).

4.5.3.2   Limitations

The main advantage of SAFIR is its robustness in modelling various modes of flexural bending behaviour of steel, concrete and composite framed structures. It does not have the capability to simulate connection behaviour. The present author has come not across any evidence to suggest SAFIR can deal with the large deflection behaviour of composite floor slabs in fire.

4.5.4   VULCAN

The computer program VULCAN is perhaps the most publicized fire-dedicated specialist program. It has been developed by successive researchers since 1985 in the Department of Civil and Structural Engineering at the University of Sheffield, UK.

The first development was carried out by Saab (1990) and Saab and Nethercot (1991). In this development, the computer program INSTAF (El-Zanaty *et al.* 1980; El-Zanaty and Murray 1980, 1983) was modified to incorporate the stress–strain relationships of steel at elevated temperatures. The Ramberg-Osgood equation was adopted to describe these stress–strain relationships.

The program INSTAF was a program originally developed by El-Zanaty and Murray at the University of Alberta, Canada to analyse the behaviour of 2D steel frames at ambient temperature. This program uses beam elements and has well-developed features to deal with non-linear material and large deflection behaviour of framed structures. Higher order terms in the strain–displacement relationships are retained so as to allow accurate treatment of large-deflection behaviour. As shown in Figure 4.6 for 2D analysis, 4 degrees of freedom are used for each node of a line element in the local direction and when transformed to global coordinates, 5 degrees of freedom are necessary. The four local degrees of freedom include the three conventional degrees of freedom of axial displacement (u), transverse displacement (v) and rotation (θ), and an additional degree of freedom u'

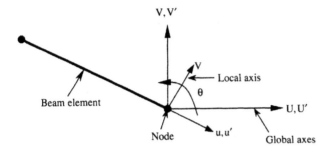

Figure 4.6 Degrees of freedom for 2D analysis in INSTAF.

(the first derivative of the axial displacement) so that the second order axial strain is included. In this program, the Total Lagrangian formulations are adopted so that moderately large deflections can be accommodated. Since the Newton–Raphson iterative method is used, it is not possible to trace the post-failure behaviour of steel frames.

Najjar (1994), Najjar and Burgess (1996) later extended the basic formulations of INSTAF for 2D analysis to three dimensions. In this program, 8 degrees of freedom are used for each node of a 2-noded line element in the local direction. These 8 degrees of freedom include the six conventional degrees of freedom (three displacements u, v, w and three rotations θ_x, θ_y, θ_z the first derivative of the axial deformation (u') and twist (θ'_z) with respect to the longitudinal axis. Inclusion of u' allows the effect of large-deflection to be considered. Inclusion of θ'_z enables simulation of warping and lateral-torsional buckling.

The next development of this program was carried out by Bailey (1995, 1998a) who introduced finite shell elements into the program so that the program could be used to analyse steel framed structures with floor slabs. Bailey's treatment of concrete slabs was essentially linear elastic and no consideration was given to high temperatures. However, the program was used to simulate one of the large-scale structural fire tests at Cardington with some success (Bailey *et al.* 1996c) and also revealed the importance of including slabs in the structure.

Bailey (1995) also added a facility to deal with semi-rigid connections. In this model, the ambient temperature approach was followed, where the connection behaviour was represented by bending moment–rotation relationships at elevated temperatures. Since the program does not have the capability to predict connection moment–rotation curves, test results such as those given in Leston-Jones (1997), Lawson (1990b) and Al-Jabri *et al.* (1998) are necessary. As discussed in Section 3.8, there are significant differences in the behaviour of a connection tested in isolation and tested as part of a complete frame. It is doubtful whether the ambient temperature approach of representing a connection will be correct in fire conditions.

The most recent development work was carried out by Huang *et al.* (1999a), who introduced a layered approach to model reinforced concrete floor slabs. In this approach, a concrete slab is divided into a number of layers in the thickness direction and reinforcements are treated as a smeared layer. Four noded shell elements are used. In order to prevent the problem of shear locking in low-order shell elements, independent shape functions for bending and shearing deformations are used (Dvorkin and Bathe 1984). The layered approach allows temperature variation in the concrete slab to be included. In the program, each layer is assumed to have a constant temperature but different layers can have different temperatures. Wang (1993, 1994b) used a similar approach in an earlier analysis of reinforced concrete slabs in fire. Huang *et al.* (2000a) further extended the capability of VULCAN to include orthotropic properties of ribbed composite slabs. This is done by using different bending stiffness in the two directions of a slab. Further development of VULCAN has made it capable of simulating membrane action in reinforced concrete floor slabs.

The VULCAN library includes line elements for frame structures and shell elements for reinforced concrete slabs. Constraint relationships may be imposed to simulate shear connectors in a composite beam (Huang *et al.* 1999b). These types of elements enable VULCAN to be used to simulate composite construction with composite floor slabs.

4.5.4.1   Applications of VULCAN

The VULCAN program has been applied to study different modes of steel structural behaviour under fire conditions, including bare steel frames in

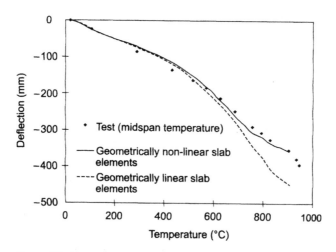

Figure 4.7 A comparison of Corus corner test and VULCAN predictions (from Huang *et al.* 2000b).

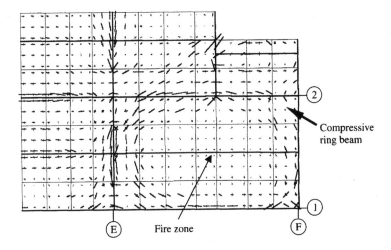

Figure 4.8 A VULCAN plot of slab stress distribution at a steel temperature of 900 °C for Corus corner fire test (from Huang et al. 2000b).

bending (Najjar and Burgess 1996), lateral torsional buckling of beams (Bailey *et al.* 1996a), composite beams with partial shear connection (Huang *et al.* 1999b), semi-rigid connection effects (Leston-Jones 1997; Bailey 1999), structural interactions (Bailey *et al.* 1996b, 1999a; Bailey 2000a,b).

However, it is without any doubt that the most extensive use of VULCAN is in simulating various Cardington fire tests (Bailey *et al.* 1996c, 1999b; Plank *et al.* 1996; Bailey 1998c; Burgess and Plank 1998; Huang *et al.* 1999c, 2000b). Figure 4.7 shows a comparison between the measured results of the Corus Corner fire test at Cardington (see Section 2.10.2.1) and VULCAN predictions. A VULCAN plot of slab stress contour in Figure 4.8 gives clear indication of tensile membrane action.

4.5.4.2 Limitations

VULCAN has only a few types of finite elements in its element library and cannot simulate many modes of detailed local structural behaviour, including local buckling, distorsional buckling and connection behaviour.

This program uses the Newton–Raphson method to perform iterations. This method is unlikely to be efficient in tracking progressive failure and load shedding.

VULCAN is a structural analysis program and temperature distributions (measured or obtained from other programs) are treated as input data.

4.5.5   Commercial programs (ABAQUS and DIANA)

ABAQUS and DIANA are commercially available general finite element programs. Although they do not have special facilities to deal with structural behaviour in fire, the effects of fire on a structure may be simulated by including relevant material properties at elevated temperatures. Both programs can be used to perform heat transfer analysis to obtain temperature distributions in structures under fire attack.

4.5.5.1   Applications of ABAQUS and DIANA

The ABAQUS program has recently been used to study the behaviour of steel and composite framed structures in fire by Edinburgh University (Gillie 1999; Gillie *et al.* 2000, 2001; Sanad *et al.* 2000a,b,c) and Corus Research in the United Kingdom (O'Callaghan and O'Connor 2000; O'Connor and Martin 1998), in connection with the Cardington research project. Ma and Makelainen also used ABAQUS to study composite frames in fire (Ma and Makelainen 1999a,b). To model composite slab behaviour in fire, Gillie (1999), Gillie *et al.* (2000), developed a user subroutine to use the stress resultant approach to deal with concrete slabs to avoid the problem of numerical non-convergence in ABAQUS when using the standard ABAQUS shell elements. Figure 4.9 shows an example of these studies, where the ABAQUS predictions are compared against the measured deflections of the restrained beam test at Cardington (cf. Section 2.10.2.1).

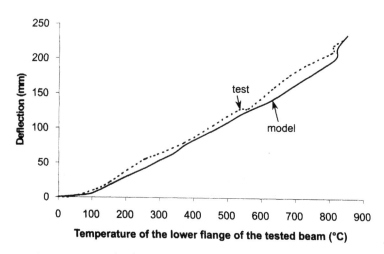

Figure 4.9 Comparison of Corus restrained beam test and ABAQUS predictions (Reprinted from the *Journal of Constructional Steel Research*, "A structural analysis of the first Cardington test". Vol. **57**, pp. 581–601, Gillie *et al.* (2001) with permission from Elsevier Science).

Evidently, ABAQUS has the ability to simulate complex structural behaviour under fire conditions. ABAQUS has also been used to study various local buckling phenomena in a steel structure (O'Connor and Martin 1998; Feng *et al.* 2001). Although the ability of ABAQUS to simulate the detailed behaviour of connections in fire has not been tested, it is expected that ABAQUS can adequately perform this task.

The DIANA program is developed at TNO, Holland and has been adapted to analyse steel framed structures under fire conditions (van Foeken and Snijder 1985). It has recently been used to simulate the Cardington fire tests (Both *et al.* 1996) with satisfactory results. It has also been extensively used to carry out detailed simulations of thermal and structural performance of composite slabs (Both *et al.* 1992, 1993, 1995, 1997; Both and Twilt 1994).

4.5.5.2 Advantages of using commercial programs

Commercial programs such as ABAQUS and DIANA have a large library of finite elements to enable efficient and detailed modelling of virtually all modes of structural behaviour involved in fire. They usually have facilities to allow users to write user defined subroutines. This is desirable and enables modelling of many of the special features of structural behaviour in fire. Being commercial programs, the advanced computational facilities in these programs to deal with material and geometrical non-linearity will have been rigorously tested by many users over a huge range of problems. They also have technical support and adequate documents about using the program.

4.5.5.3 Limitations

Despite the huge advantages of using commercial programs, they do have some disadvantages. These general finite element analysis programs usually require substantial initial investment for buying the software and associated hardware, and training up a specialist analyst. They also need costly maintenance for renewing the software license and retaining the specialist analyst. Because of these problems, commercial programs are not as portable as specialist ones. If the user's interest is only in structural behaviour in fire, it is perhaps more cost-effective to have a specialist fire dedicated program.

4.6 SIMPLIFIED FRAME ANALYSIS METHODS

Computer programs based on finite element analysis have the advantage of being general and versatile and being able to give detailed information, such as deformation and stress histories of any part of a structure under fire conditions. However, these numerical tools generally require highly skilled expertise to use and can be laborious when setting up a model for simulation.

Furthermore, in structural fire safety analysis, if the ultimate load or the ultimate temperature is the main concern, detailed deformation or stress history is only of secondary importance. In order to make structural fire analysis more accessible, it is desirable to develop more straightforward and simpler analytical tools.

Current design procedures represent the simplest method, but they are based on elemental structural behaviour and cannot deal with many factors in framed structural behaviour where there are redistributions of load and interactions between different members. There is now a tendency to develop simplified analytical methods that can deal with the major aspects of complete frame behaviour and at the same time, can also be easily used by practising engineers. Two types of simplification may be employed. The first uses repeated analysis to obtain force distributions and remove members that have failed. The second uses plastic analysis, where non-linear material behaviour is considered but stability effects are ignored.

Examples of simplified frame analysis in fire include the work of Liew *et al.* (1998), Li and Jiang (1999), Chan and Chan (2001) which is based on the computer program USFOS (Amdahl and Hellan 1992; Amdahl *et al.* 1995; Tan 2000; Wong 2001).

Of course, these "simpler" methods all have a number of limitations. They usually have to make simplifications about temperature distributions in the structure under fire attack. All programs use conventional 6 degrees of freedom per node for 3D analysis (3 translations and 3 rotations) or 3 degrees of freedom per node (2 displacements and one rotation) for 2D analysis. Therefore, they cannot deal with local, distortional or lateral torsional buckling. They may be able to deal with semi-rigid connection behaviour only if the connection behaviour is represented by moment–rotational relationships. They cannot be used to analyse detailed connection behaviour. None of them has the capability to deal with the influence of concrete floor slabs and composite construction.

4.7 SUMMARY AND SOME RECOMMENDATIONS

The information in this chapter is based on those published by the various program developers. Without personal use of all these computer programs, it is difficult to give a thorough assessment of the capability, accuracy and user-friendliness of these computer programs. Nevertheless, it is possible to make some general suggestions about the suitability of these computer programs to deal with various aspects of steel structural behaviour in fire and to point out some ways forward to enhance the capability of these computer programs. Tables 4.1 and 4.2 presents a summary of the assessment.

Clearly, the development of computer programs to analyse the behaviour of steel and composite structures in fire has made tremendous progress in

Table 4.1 A summary of the reviewed programs for modelling local behaviour

Name	Bending/shear		Local/ distorsional buckling	Lateral torsional buckling	Composite slab/ Membrane action	Detailed connection modelling
	Steel	Composite				
ADAPTIC	✓	✗	✗	✗	✓	✗
FEAST	✓	✗	✓	✓	✗	✓
SAFIR	✓	✓	✓	✓	✗	✗
VULCAN	✓	✓	✗	✓	✓	✗
ABAQUS	✓	✓	✓	✓	✓	✓
DIANA	✓	✓	✓	✓	✓	✓

Table 4.2 A summary of the reviewed programs for modelling global behaviour

Name	Thermal analysis	Frame analysis	Non-linear material	Large deflections	Progressive failure
ADAPTIC	✗	✓	✓	✓	✓
FEAST	✗	✗	✓	✓	✗
SAFIR	✓	✓	✓	✓	✗
VULCAN	✗	✓	✓	✓	✗
ABAQUS	✓	✓	✓	✓	✓
DIANA	✓	✓	✓	✓	✓

the last 10 years or so. Nevertheless, there is still some scope for each program to be further developed and tested so that they can be confidently used to model all aspects of structural behaviour in fire with accuracy and efficiency.

Chapter 5

Behaviour of steel and composite structures in fire

Chapter 3 has provided a qualitative description of the various modes of behaviour of steel structure in fire, based on experimental observations. Even though test results can sometimes be directly used in general design, this is rare and the purpose of carrying out fire tests is far more likely to provide the basis for further detailed studies. It is through detailed analyses of test results that a thoroughly quantitative understanding is developed so that better design guidance can be produced. This chapter will present a systematic description of the results of a variety of quantitative studies of the different modes of behaviour of steel structures in fire. This knowledge is necessary in order to understand current design provisions that will be presented in Chapters 8, 9 and 10, and also to appreciate how further design methods may incorporate new understandings. Due to the importance of interactions between different structural members, this chapter will emphasize on the behaviour of beams and columns that form part of a complete building structure.

5.1 MATERIAL PROPERTIES AT ELEVATED TEMPERATURES

The structural effects of a fire on the behaviour of a steel structure are caused by:

- Changes in the mechanical properties of steel and concrete. Both materials become weaker and more flexible at high temperatures.
- Temperature induced strains.

These changes lead to various phenomena observed in different fire tests. Therefore, to understand the complex behaviour of a steel structure under fire conditions, it is necessary to avail the basic information of material properties at elevated temperatures. They include the stress–strain relationships of steel and concrete at elevated temperatures, reductions in the strength and stiffness of steel and concrete at elevated temperatures and their thermal strains. Thermal properties that are required for temperature analysis will be given in Chapter 6.

Descriptions of material properties will mainly follow the guidance given in Eurocode 3 Part 1.2 (CEN 2000b) and Eurocode 4 Part 1.2 (CEN 2001). When the ranges of application of these Eurocodes are exceeded, additional information from other relevant sources will be provided.

5.1.1   Steel

5.1.1.1   Temperature induced strains

The temperature-induced strains of steel include thermal expansion and creep strain.

Thermal expansion of steel

The thermal expansion strain (ε_{th}) of steel is given by

$$
\begin{aligned}
\varepsilon_{th} &= -2.416 \times 10^{-4} + 1.2 \times 10^{-5}T + 0.4 \times 10^{-8}T^2 &&\text{for} \quad T \leq 750\,^\circ\text{C}; \\
\varepsilon_{th} &= 0.011 &&\text{for} \quad 750 < T \leq 860\,^\circ\text{C}; \\
\varepsilon_{th} &= -0.0062 + 2 \times 10^{-5}T &&\text{for} \quad T > 860\,^\circ\text{C}
\end{aligned}
$$

$$(5.1)$$

A graphical representation of these relationships is shown in Figure 5.1. The step change in the temperature range 750–860 °C is due to a phase change in the steel.

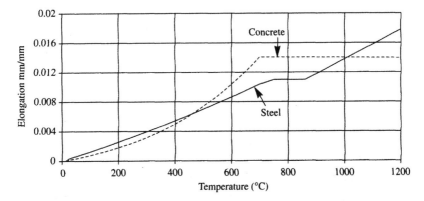

Figure 5.1 Thermal elongation of steel and concrete as a function of temperature (from CEN 2001). Reproduced with the permission of the British Standards Institution under licence number 2001SK/0298.

In simple calculations, the coefficient of thermal expansion of steel may be assumed to be a constant so that the thermal expansion strain is given by:

$$\varepsilon_{th} = 14 \times 10^{-5}\Delta T \qquad (5.2)$$

Creep strain

The high temperature creep strain of steel expresses the increase in the steel strain when steel is simultaneously subjected to high temperature and high stress over time. Results of various creep strain tests show that the steel creep strain consists of three parts: primary, secondary and tertiary creep strains as illustrated in Figure 5.2. Due to the relatively short exposure time of fire attack, only the primary and secondary creep strains need be considered. A simplified creep strain model of steel is that of Plen (1975), based on the Dorn–Harmathy creep strain model (Dorn 1954; Harmathy 1967). In this simplified model, the primary creep strain is described by a parabola and the secondary creep strain by a straight line, giving:

$$\varepsilon_{cr} = \varepsilon_{cr0}\left(2\sqrt{\frac{Z \times \theta}{\varepsilon_{cr0}}}\right) \quad \text{for} \quad \theta \leq \theta_0;$$

$$\varepsilon_{cr} = \varepsilon_{cr0} + Z \times \theta \qquad \text{for} \quad \theta > \theta_0 \qquad (5.3)$$

in equation (5.3) θ is the temperature compensated time, Z is the slope of the secondary creep strain–time relationship and ε_{cr0} is the intercept at the creep

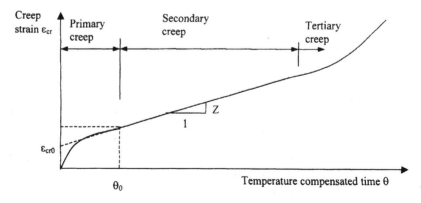

Figure 5.2 Creep strain model for steel at high temperatures.

strain axis by the secondary creep strain slope line. The temperature compensated time is evaluated according to the following expression:

$$\theta = \int_0^t e^{-\frac{\Delta H}{RT}} dt \tag{5.4}$$

where t is time in seconds, T is the temperature in K and $\Delta H/R$ is a constant.

From equation (5.3), the temperature compensated time (θ_0) that divides the primary creep and the secondary creep strains can be obtained as:

$$\theta_0 = \frac{\varepsilon_{cr0}}{Z} \tag{5.5}$$

Detailed calculations of steel creep strain at high temperatures are tedious. Fortunately, for practical considerations of steel structures under fire conditions, the period of time when a steel structure is exposed to high temperature is short so that the creep strain may be neglected. To compensate for this simplification, the effect of creep strain is implicitly included in the stress–strain relationships of steel at high temperatures.

Consider the following example. Grade S275 steel has similar properties to those of ASTM A36 to be used as input data in equations (5.3) and (5.4). Using data from Anderberg (1983): $\Delta H/R = 70\,000$ K, $\varepsilon_{cr0}(\sigma) = 1.7 \times 10^{-10}\sigma^{1.75}$ and $Z(\sigma) = 1.23 \times 10^{16}e^{0.0003\sigma}$ for $15\,000 < \sigma < 45\,000$ psi (103–310 N/mm^2). Considering a constant steel temperature of 1200 K (927 °C) with a constant stress of 200 N/mm^2 (28 996 psi) for a time duration of 1 h, gives:

$$\theta = 5.6361 \times 10^{-26}\,\text{s}, \quad \varepsilon_{cr0} = 0.01095, \quad Z = 7.3747 \times 10^{19}$$
$$\text{and} \quad \theta_0 = 1.4848 \times 10^{-22}$$

From equation (5.3), $\varepsilon_{cr} = 0.00039$. Even for this combination of high temperature, high stress level and long duration, this level of creep strain is very low. In fact such a stress level could not be maintained at such a high temperature and the stress level, temperature and time duration used in the example are all near the maximum value ever to be found in practice. In more realistic combinations of these properties, the creep strain will be smaller by orders of magnitude. This justifies the practice of neglecting creep strain in more detailed numerical analyses.

5.1.1.2 Stress–strain relationships of steel at elevated temperatures

Conventional hot-rolled structural steels

The stress–strain relationships of steel at elevated temperatures depend on whether steady state or transient state testing is employed. Transient state testing is preferred since it reflects the realistic situation that for a structure in fire, mechanical loading is applied before fire exposure. In transient state testing, the rate of heating has some influence due to creep strain. But since the steel creep strain is small, as demonstrated above, mechanical testing of steel at elevated temperatures is usually carried out by using a typical heating rate of about 10 °C/min as found in realistic steel structures exposed to fire conditions. Although a variety of mathematical models have been proposed for the stress–strain relationships of steel at elevated temperatures, they are based on the same set of test results from Corus (Kirby and Preston 1988). The model in Eurocodes 3 and 4 Part 1.2 (CEN 2000b, 2001) is well accepted and will be described here.

In the Eurocodes, the stress–strain curve of steel consists of a straight line for the initial response, followed by an elliptical relationship and then a plateau. Table 5.1 gives the mathematical descriptions used in the Eurocodes and Figure 5.3 provides an illustration of this model and shows various parameters to be used in the mathematical model. In order to use this model, the reduced strength and stiffness of steel at elevated temperatures are required as input data and Table 5.2 gives their values, expressed as ratios of the value at elevated temperature to that at ambient temperature. These ratios are often referred to as retention factors. A graphic representation of these retention factors is shown in Figure 5.4.

Cold-formed thin-walled steel

During the cold-forming process, steel may be stressed beyond its yield strength into strain hardening. Because of this, cold-formed steel may have a higher effective yield strength than hot-rolled steel. However, this increase in

Table 5.1 Mathematical model of the stress–strain relationship of steel at elevated temperatures

Strain range	stress σ
$\varepsilon \leq \varepsilon_{p,T}$	εE_T
$\varepsilon_{p,T} < \varepsilon < \varepsilon_{y,T}$	$p_{p,T} - c + \dfrac{b}{a}\sqrt{[a^2 - (\varepsilon_{y,T} - \varepsilon)^2]}$
$\varepsilon_{y,T} \leq \varepsilon \leq \varepsilon_{t,T}$	$p_{y,T}$
Parameters	$\varepsilon_{p,T} = \dfrac{p_{p,T}}{E_T}, \; \varepsilon_{y,T} = 0.02, \varepsilon_{t,T} = 0.15$
Functions	$a^2 = (\varepsilon_{y,T} - \varepsilon_{p,T})(\varepsilon_{y,T} - \varepsilon_{p,T} + \frac{c}{E_T})$
	$b^2 = c(\varepsilon_{y,T} - \varepsilon_{p,T})E_T + c^2$
	$c = \dfrac{(p_{y,T} - p_{p,T})^2}{(\varepsilon_{y,T} - \varepsilon_{p,T})E_T - 2(p_{y,T} - p_{p,T})}$

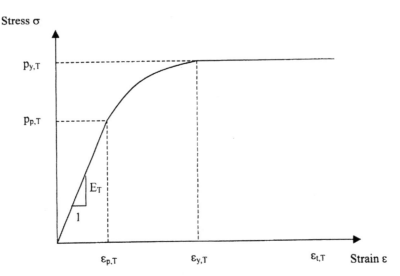

Figure 5.3 Stress–strain relationship of hot-rolled steel at elevated temperatures.

Table 5.2 Retention factors of steel at elevated temperatures

Steel temperature T (°C)	Effective yield strength (relative to p_y at 20°C) $k_{y,T} = p_{y,T}/p_y$	Proportional limit (relative to p_y at 20°C) $k_{p,T} = p_{p,T}/p_y$	Slope of the linear elastic range (relative to E_a at 20°C) $k_{E,T} = E_T/E_a$
20	1	1	1
100	1	1	1
200	1	0.807	0.9
300	1	0.613	0.8
400	1	0.42	0.7
500	0.78	0.36	0.6
600	0.47	0.18	0.31
700	0.23	0.075	0.13
800	0.11	0.050	0.09
900	0.06	0.0375	0.0675
1000	0.04	0.025	0.045
1100	0.02	0.0125	0.0225
1200	0	0	0

Source: CEN (2000b). Reproduced with the permission of the British Standards Institution under licence number 2001SK/0298.

Note

The effective yield strength is defined at 2% strain.

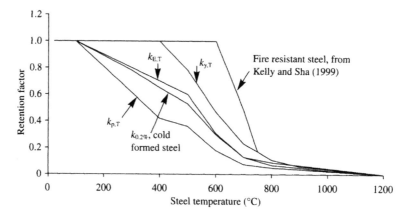

Figure 5.4 Retention factors of steel at elevated temperatures (from CEN 2000b). Reproduced with the permission of the British Standards Institution under licence number 2001SK/0298.

the steel yield strength by cold-forming is gradually reduced at elevated temperatures. Therefore, the retention factors of cold-formed steel are lower than those of hot-rolled steel even though the absolute values may be higher. Another difference between cold-formed thin-walled steel and hot-rolled steel is the definition of yield strength. In the Eurocodes for hot-rolled steels (S275 and S355), the effective yield strength is taken at 2% strain. For cold-formed thin-walled steel where local buckling has great influence on design, the effective yield strength is defined as that at 0.2% plastic strain (or 0.2% proof stress). Figure 5.4 compares the effective yield strengths of hot-rolled steel and cold-formed steel according to Eurocode 3 Part 1.2 (CEN 2000b). Other sources of information on the mechanical properties of cold-formed steel are provided by Outinen *et al.* (2000), Outinen and Makelainen (2001) who carried out extensive coupon tests at elevated temperatures and gave retention factors at 2% strain, and Lawson (1993) who gave retention factors at 0.5% and 1.5% strain levels.

Fire resistant steel

The strength of steel at elevated temperatures may be improved by adding alloy elements to make steel more fire resistant. Sakumoto *et al.* (1993, 1994), Sakumoto (1995, 1998) have presented the mechanical properties of one type of FR steel used in Japan. It is manufactured by adding molybdenum (Mo) and other alloying elements to conventional structural steel. These alloying elements are converted at high temperatures, to compounds that dissolve into an atomic arrangement and these alloying elements suppress shearing among atoms, thus producing high resistance.

The change in the Young's modulus of fire resistant steel is similar to that of the conventional steel at elevated temperatures. Figure 5.4 compares the strength retention factors of a FR steel (Kelly and Sha 1999) with those of a conventional steel. It can be seen that the FR steel maintains a much higher strength at temperatures of around 600 °C (often regarded to be the critical temperature of steel). At higher steel temperatures (> 750 °C), the strength retention factors of both types of steel are similar.

5.1.2 Concrete

5.1.2.1 Thermal strains

The thermal strain of concrete is complex and is influenced by a number of factors. According to Anderberg and Thelandersson (1976); Khoury *et al.* (1985); Khoury (1983), the thermal strain of concrete may be divided into thermal expansion strain, creep strain and a stress induced transient thermal strain. Interested readers should refer to the above references for more detailed information on how to evaluate these different strain components. Eurocode 4 Part 1.2 (CEN 2001) takes a simple approach and gives the coefficient of thermal expansion of concrete as:

$$\varepsilon_{th} = -1.8 \times 10^{-4} + 9 \times 10^{-6}T + 2.3 \times 10^{-11}T^3 \quad \text{for} \quad 20\,°C \leq T \leq 700\,°C;$$
$$\varepsilon_{th} = 0.0014 \qquad\qquad\qquad\qquad\qquad\quad \text{for} \quad T > 700\,°C$$

$$(5.6)$$

Figure 5.1 compares the thermal expansion strains of steel and concrete.

5.1.2.2 Stress–strain relationships

The mechanical properties of concrete are more variable than those of steel. Phan and Carino (1998, 2000) recently carried out a survey of mechanical properties of concrete (including high strength concrete) at elevated temperatures. Khoury (1992) provided an explanation of the variability in concrete mechanical properties at elevated temperatures. Values in Eurocode 4 Part 1.2 (CEN 2001) may be regarded as the lower bound values of different test results for normal strength concrete.

Figure 5.5 shows the Eurocode 4 model for the stress–strain relationship of concrete and definitions of various parameters. The stress–strain relationship is divided into two parts: the ascending part and the descending part.

The Eurocode 4 equation for the ascending part is:

$$\sigma_{c,T} = f_{c,T}\left\{3\left(\frac{\varepsilon_{c,T}}{\varepsilon_{cu,T}}\right)\Bigg/\left[2 + \left(\frac{\varepsilon_{c,T}}{\varepsilon_{cu,T}}\right)^3\right]\right\}$$

$$(5.7)$$

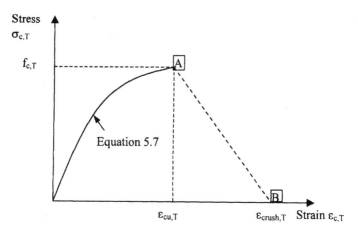

Figure 5.5 Stress–strain relationship of concrete at elevated temperatures.

where $\sigma_{c,T}$, $\varepsilon_{c,T}$, $f_{c,T}$, $\varepsilon_{cu,T}$ are respectively the stress, strain, peak stress and strain at peak stress for concrete at elevated temperature T.

From equation (5.7), the initial Young's modulus of concrete may be obtained from:

$$E_{c,T} = \frac{3}{2} \frac{f_{c,T}}{\varepsilon_{cu,T}}$$

(5.8)

Table 5.3 Strength and strain limits of normal weight concrete (NWC) and lightweight concrete (LWT) at elevated temperatures

Temperature	$k_{c,T} = f_{c,T}/f_{c,a}$		$\varepsilon_{cu,T} \times 10^3$	$\varepsilon_{crush,T} \times 10^3$
	NWC	LWC		
20	1	1	2.5	20.0
100	0.95	1	3.5	22.5
200	0.9	1	4.5	25.0
300	0.85	1	6.0	27.5
400	0.75	0.88	7.5	30.0
500	0.60	0.76	9.5	32.5
600	0.45	0.64	12.5	35.0
700	0.30	0.52	14.0	37.5
800	0.15	0.4	14.5	40.0
900	0.08	0.28	15.0	42.5
1000	0.04	0.16	15.0	45.0
1100	0.01	0.04	15.0	47.5
1200	0.0	0.0	15.0	50.0

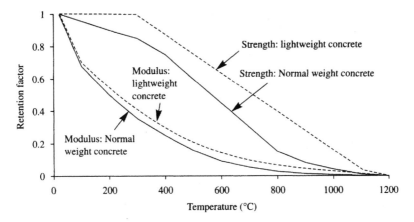

Figure 5.6 Retention factors of concrete at elevated temperatures (from CEN 2001). Reproduced with the permission of the British Standards Institution under licence number 2001SK/0298.

The descending part is a straight line, joining the peak point (A) with the point of concrete crush (B) in Figure 5.5.

Values of $f_{c,T}$, $\varepsilon_{cu,T}$ and $\varepsilon_{crush,T}$ are required to determine the complete stress–strain relationship of concrete at elevated temperatures. Table 5.3 gives their values recommended by Eurocode 4 Part 1.2 (CEN 2001).

Figure 5.6 shows the retention factors for the strength ($f_{c,T}$) and stiffness ($E_{c,T}$ from equation (5.8)) for normal weight and lightweight concretes at elevated temperatures.

5.2   BEHAVIOUR OF UNRESTRAINED COLUMNS IN FIRE

Columns are structural members under predominantly compression load and are susceptible to local buckling, global buckling and cross-sectional yield. Local buckling is dealt with separately, while global buckling and cross-sectional yield are combined together. In this section, columns with uniform and non-uniform temperature distributions are discussed separately. Whilst the behaviour of a column with uniform temperature distribution is relatively simple, the behaviour of a column with non-uniform temperature distributions is complex due to thermal bowing induced additional bending moments and thermal stresses.

5.2.1   Local buckling

In the design of a steel structure where local buckling may occur, it is convenient to use the concept of "effective width" to evaluate the load carrying capacity of a steel plate under compression. The effective width may be defined as the width of the plate that sustains the maximum stress in uniform compression such that the total load sustained by the effective width is the same as that by the plate. The concept of effective width is illustrated in Figure 2.2 in Chapter 2. Failure of the steel plate occurs when the uniform stress acting on the effective width has reached yield.

The effective width method underpins the design of thin-walled steel plates under compression. The effective width of a steel plate depends on a number of factors such as the width to thickness ratio, the support condition and the stress distribution in the plate. For a comprehensive review and derivations of effective widths for various situations at ambient temperature text books such as Galambos (1998), Rhodes (1991) and Yu (1991) should be consulted.

At elevated temperatures, the results of a study by Ranby (1998) indicate that the "effective width" method may be extended to the fire situation. This has also been suggested in a separate study by Uy and Bradford (1995). In the study of Ranby, analyses were carried out to determine the effective width of steel plates with width to thickness ratios of 100, 67, 50 and 40 at different elevated temperatures. Provided the temperature distribution is uniform, the same ambient temperature design equations for "effective width" may be used at elevated temperatures. Obviously, the yield stress and Young's modulus of steel at ambient temperature should be replaced by those at elevated temperatures.

Since the stress–strain relationships of steel at elevated temperatures are highly non-linear, the problem of applying the ambient temperature method for local buckling becomes how to determine the appropriate level of the yield stress of steel at elevated temperatures. Ranby (1998) suggests that either 0.1% or 0.2% proof stress may be used. For cold-formed thin-walled steel, Eurocode 3 Part 1.2 (CEN 2000b) gives the 0.2% proof stress and the retention factors are shown in Figure 5.4. For hot-rolled conventional structural steels, the 0.2% stress may be obtained from the equations in Table 5.1. Figure 5.7 shows that the 0.2% proof stress of hot-rolled steels have similar retention factors to those of cold-formed steel, and that the 0.2% proof stress may be obtained as the average of the proportional limit and the effective yield strength at 2% strain.

If the steel plate has non-uniform temperature and stress distributions, the local buckling problem becomes complex. However, it may be seen from Figure 5.7 that the reductions in the 0.2% proof stress and the Young's modulus of steel are similar so that the ratio of the Young's modulus to the 0.2% proof stress is relatively unchanged at elevated temperatures. The effective width is related to the square root of this ratio and it can be seen from Figure 5.7 that this value is very close to 1.0 until a steel temperature of about 600 °C. Even at higher temperatures, this value is less than 1.1. Hence the effective width will only change very slightly within a wide range of realistic steel temperatures. For simplicity, the effective width of a steel plate at

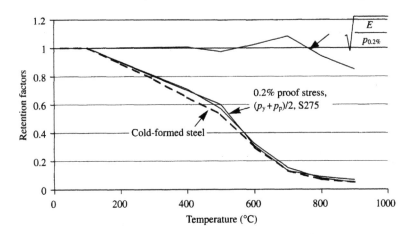

Figure 5.7 Retention factors for local buckling (from CEN 2000b). Reproduced with the permission of the British Standards Institution under licence number 2001SK/0298.

elevated temperatures may be taken as that at ambient temperature. In any case, the retention factor for Young's modulus only starts to reduce faster than that for the 0.2% proof stress at temperatures exceeding 800 °C. It would be safe to use the ambient temperature effective width for fire design.

5.2.2   Global behaviour

5.2.2.1   Columns with uniform temperature distribution

Figure 5.8 shows the typical temperature–axial deformation behaviour of a steel column subjected to the standard fire exposure on all sides. Details of this column are taken from test 41 of Wainman and Kirby (1987). The applied stress on the column was 138 N/mm² and the column length was 3000 mm. Also plotted in Figure 5.8 are the free thermal expansion of the column and the difference between the free thermal deformation and the recorded total deformation, which may be taken as the recorded mechanical deflection of the column. The calculated mechanical deformation of the column shown in Figure 5.8 was obtained from the measured stress–strain relationships of steel at elevated temperatures (Kirby and Preston 1988) at an applied stress of 138 N/mm². It can be seen from Figure 5.8 that the column axial deformation was mainly due to the free thermal expansion at elevated temperatures. The mechanical deformation caused by the applied load was small, even at relatively high temperatures. This behaviour lasted until the column approached failure when the column axial stiffness became very low and the accelerated axial shortening under load at high temperatures caused the column axial

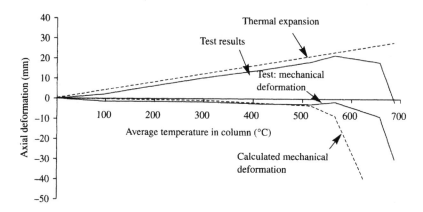

Figure 5.8 Axial deformation–time relationship of a steel column in fire (from Wainman and Kirby 1987, Test 41). Reproduced with the permission of Corus.

deformation to reverse in direction. If the thermal elongation of the column is subtracted from the total column deformation, the remaining column deformation is easily explained by the increase in column contraction due to reducing column stiffness at increasing temperatures. The effect of thermal expansion is simply to cause the column to expand without any additional mechanical effects. Column failure occurred at a temperature at which the applied stress (138 N/mm^2) exceeded the steel strength, i.e. the column failure mode was complete yield of steel in compression.

Thus, the global behaviour of a uniformly heated steel column without any restraint is simple, and is identical to that of a column at ambient temperature. When evaluating the column behaviour when heated, the only additional consideration is that the reduced strength and stiffness of steel at elevated temperatures should be used.

5.2.2.2 Effects of non-uniform temperature distributions

A steel column may experience non-uniform heating, both in the cross-section and along the length of the column. Examples of non-uniform heating in the cross-section include columns with blocked-in webs, shielded flanges, fire exposure on one side or external fire exposure. Non-uniform temperature distribution in the longitudinal direction of a column may be caused by non-uniform fire exposure, e.g. localized heating.

Effects of non-uniform temperature distribution in the cross-section

Non-uniform temperature distribution in the cross-section of a column can affect the column behaviour in a variety of ways and it is misleading to use the average column temperature to assess the column behaviour in fire. Non-uniform temperature distributions can introduce additional bending moments in the column due to thermal bowing and this can reduce the column capacity. In addition, in a cross-section with non-linear temperature distribution causing non-linear thermal strain distribution, in order for a plane section to remain planar, some parts of the column cross-section may become highly stressed and lose stiffness even without any applied load. This can give the column a reduced buckling strength compared with that of a column with uniform-temperature distribution. Thus, depending on the column failure mode and the column temperature distribution, a column with non-uniform temperature distribution may have a higher or lower maximum failure temperature than that with uniform temperature distribution. The results of a numerical study by Wang *et al.* (1995) may be used to illustrate this effect.

Consider the behaviour of a column with different types of non-uniform temperature distribution shown in Figure 5.9. Type 1 is the basic uniform heating. Types 2 and 3 have a linear temperature distribution with different gradients representing two possible extremes of a column heated from one side. Type 4 approximates a column with blocked-in web and type 5 represents a column constructed in wall, therefore only one flange is heated. Types 2, 3 and 5 give unsymmetrical temperature distributions, therefore producing thermal bowing and additional bending moments. Type 4 has a symmetrical temperature distribution and there is no thermal bowing induced bending moment.

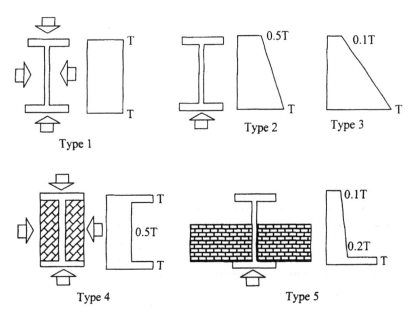

Figure 5.9 Different types of assumed non-uniform temperature distribution (from Wang et al. 1995).

Table 5.4 Failure temperatures of columns with non-uniform temperature distributions

Load ratio	Heating type	Column failure temperature (°C) at a slenderness of					
		20	40	60	80	100	150
	1	**656**	**598**	**586**	**586**	**594**	**670**
	2	690	670	728	616	548	486
0.3	3	722	748	730	574	490	426
	4	712	682	588	534	512	520
	5	742	742	818	574	472	422
	1	**580**	**514**	**474**	**476**	**492**	**550**
	2	596	570	480	406	366	320
0.5	3	622	626	480	318	260	222
	4	596	462	440	389	403	432
	5	625	615	367	261	223	198
	1	**494**	**380**	**350**	**380**	**420**	**444**
	2	512	454	288	226	194	178
0.7	3	550	548	210	144	120	104
	4	524	436	240	250	284	347
	5	544	347	170	126	108	96

Source: Wang et al. (1995).

Wang et al. (1995) carried out numerical analyses for different column slenderness and levels of loading. In all analyses, the column section was 203 × 203 × 52 UC and grade S275 steel was used. Analytical results are summarized in Table 5.4, showing the maximum column failure temperature for load ratios of 0.3, 0.5 and 0.7 and for column slenderness ranging from 20 to 150. The load ratio is the ratio of the applied load to the column load carrying capacity at ambient temperature. Failure temperatures for uniform temperature distribution are given in bold.

From the results in Table 5.4, a number of observations may be made with regard to the behaviour of columns with non-uniform temperature distributions.

- For a column of low slenderness (<40), since the column failure mode is cross-sectional yielding, the average temperature of the column is the most important parameter. Thus, a column with non-uniform temperature distribution will have a higher maximum flange temperature at failure than a uniformly heated column.
- For a column of high slenderness, the maximum column temperature at failure is lower in the case of non-uniform temperature distribution. There are three contributory factors. First, non-uniform temperature distribution gives a non-uniform distribution of material strength and

stiffness, causing a shift in the centroid of the cross-section which generates an eccentricity for the axial load. Second, in order for the total strain in the cross-section to remain linear, additional mechanical strains are produced if the thermal strain distribution is non-linear. These additional strains cause a reduction in the tangent stiffness of the cross-section, leading to early column failure. Third, non-linear temperature distribution can introduce thermal bowing, which generates additional bending moments due to the P-δ effect.

• At high load ratios, because the applied column stress is high, the column failure temperature can be severely reduced due to non-uniform temperature distribution. For example, at a load ratio of 0.7, for a range of practical column slenderness (>60), the column failure temperature may fall under 200 °C. Clearly, the designer should carefully consider the use of slender columns in situations where non-uniform temperatures may occur.

Other investigators (Burgess *et al.* 1992; Shepherd 1999) have also reached similar conclusions with regard to failure temperatures of columns with non-uniform temperature distributions. The design implication is that the designer should not automatically assume that because a column is non-uniformly heated so that it has a low average temperature, it should be able to survive fire attack for a longer period. This also raises doubt as to whether fire resistance tests on columns with uniform temperature distribution can be safely extrapolated to columns with non-uniform temperature distributions. At present, this problem has not been satisfactorily resolved in simple design methods and advanced finite element analysis may become necessary.

Effects of non-uniform temperature distribution along the column length

When a column is tested in a standard fire testing furnace, the typical temperature distribution in the column is of low temperatures at the ends and high temperatures in the middle. This appears to be because the column ends are in close proximity to the cold ceiling and floor. Witteveen and Twilt (1981–82) examined the influence of this type of temperature distribution on failure temperatures of steel columns. If the maximum failure temperatures in the column mid-height for different non-uniform temperature distributions are compared, they found that depending on the column slenderness, the column end conditions and applied load, the increase in the column failure temperature due to non-uniform heating along the column height could be substantial. For example, as shown in Figure 5.10, the difference in column failure temperatures between a uniform distribution and a triangular distribution can be higher than 100 °C. Fortunately, the effect of non-uniform heating along the column length is to give the column a higher failure temperature than under uniform heating. Therefore, it is safe if this effect is ignored.

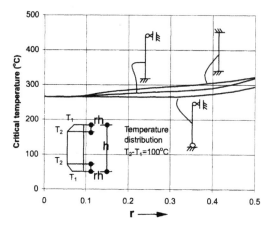

Figure 5.10 Effect of longitudinal non-uniform temperature distribution on column failure temperature (Reprinted from the *Fire Safety Journal*, "A critical view on the results of standard fire resistance tests on steel columns". Vol. 4, pp. 259–270, Witteveen, J. and Twilt, L. (1981/82) with permission from Elsevier Science).

5.2.3 Composite columns

5.2.3.1 Local buckling

The need to consider local buckling in a composite column is greatly reduced for the reasons listed below:

- If steel is encased in concrete, local buckling will not occur.
- In a concrete filled tube, if fire protection is by board spray, it is likely that local-buckling of the thin-walled steel plates would be restrained by the fire protection. If the tube is protected by an intumescent coating, local buckling may occur and should be considered. The effective width at ambient temperature may be used for fire design.
- In a concrete filled tube without fire protection, due to very high temperatures attained in the steel tube, contributions from the steel tube will be minimal and may be neglected. Hence whether or not the steel will experience local buckling will not be important.

5.2.3.2 Global behaviour

In Section 3.2, it was shown that the deformation behaviour of a composite column could be complex due to load transfer from steel to concrete. However, as will be shown in Section 9.3, calculations of the strength for steel and composite columns are identical.

5.2.3.3 Confinement effect

At ambient temperature, it is possible, in a short concrete filled circular hollow section, to utilize the confinement effect from the steel tube on the concrete filling to increase the squash resistance of the column. The confinement effect depends on the steel strength and the rate of concrete dilation approaching failure. Under fire conditions, the confinement effect may not be reliable for the following two reasons, both related to the fact that the steel tube is at a much higher temperature than the concrete core:

1   The strength of the steel is reduced much more than the concrete;
2   The steel tube has a much higher lateral expansion than the concrete core, which relieves the confinement effect on concrete dilation.

5.3   BEHAVIOUR OF RESTRAINED COLUMNS IN FIRE

When a column forms part of a statically indeterminate structure, it will have many interactions with the adjacent structure. These interactions become complicated under fire conditions. Possible effects of structural interaction on the behaviour of a column in fire include:

●   Change in the column axial load due to axial restraint to its thermal expansion as shown in Figure 5.11;
●   Change in the column buckling length due to variable rotational restraint to the column;
●   Change in the column bending moment due to variable bending stiffness of the column relative to the adjacent structure;
●   Change in the column bending moment due to lateral load of the thermally restrained adjacent beam as shown in Figure 5.12;
●   Change in the column bending moment due to the column axial load acting on the thermal expansion of the adjacent beam as shown in Figure 5.13.

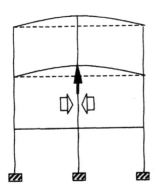

Figure 5.11 Effect of restrained thermal expansion.

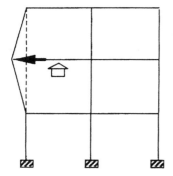

Figure 5.12 Effect of push over by the adjacent beam.

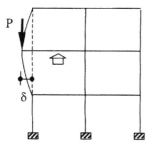

Figure 5.13 P-δ effect.

All these interactions should be considered so that the loading and boundary conditions of the column can be precisely determined for design calculations.

5.3.1   Effects of axial restraint on column thermal expansion

When the axial deformation of a column is different from those of others on the same floor, restraints from the floors above the column will induce an additional axial load in the column. Differential axial deformation of the column may be a result of either differential thermal expansions or different rates of mechanical deformation or both. Under fire conditions, differential thermal expansions are the main cause of differential column deformations.

This effect can be represented by a column with a restraining spring attached at one end, as shown in Figure 5.14. The spring has an axial stiffness of K_s to represent the restraint stiffness of the adjacent structure.

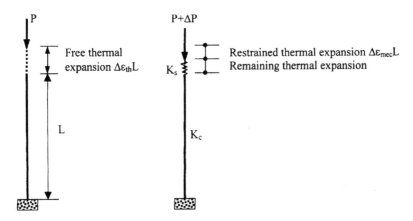

Figure 5.14 Analytical model for an axially restrained column.

Consider a uniformly heated column under axial load. After a temperature rise ΔT, an additional compressive load (ΔP) is generated in the column and this may be obtained from the following equation:

$$\Delta P = \frac{K_c K_s}{K_c + K_s}(\Delta\varepsilon_{th} - \Delta\varepsilon_{mec})L \qquad (5.9)$$

where K_c is the tangent axial stiffness of the column at elevated temperature, $\Delta\varepsilon_{th}$ and $\Delta\varepsilon_{mec}$ are respectively increments in the free thermal and compressive mechanical strains of the column during a temperature rise ΔT and L is the column length. If $\Delta\varepsilon_{th}$ is greater than $\Delta\varepsilon_{mec}$, an additional compressive load is generated in the column. Otherwise, the column compressive load is reduced. In some early attempts to study the behaviour of an axially restrained column at elevated temperatures, the column mechanical strain was not included. As a result, very high values of additional column compressive load were obtained.

Since the tangent stiffness of the heated column is temperature dependent, it is better to use the initial column stiffness to characterize the degree of restraint by the adjacent structure. Thus, equation (5.9) may be re-written as:

$$\Delta P = \frac{RK_c}{K_c/K_{c0} + R}(\Delta\varepsilon_{th} - \Delta\varepsilon_{mec})L \qquad (5.10)$$

where $R = K_s/K_{c0}$ is the ratio of the restraint stiffness to the initial column stiffness K_{c0}. According to Wang and Moore (1994), the value of R in practical structures is about 2%.

Wang (1994a), Wang and Moore (1994) checked the accuracy of equation (5.10) by comparing predicted results using equation (5.10) against those of a finite element study. Simms *et al.* (1995–96) also used equation (5.9) and checked its predictions against the results of a series of experimental studies they have carried out and confirmed the accuracy of this equation (cf. Figure 3.7).

As mentioned in Section 5.2.2, the axial deformation in a heated column is largely thermal expansion with a relatively small mechanical contraction that occurs near column failure. Therefore, it is expected that equation (5.10) will usually produce an additional compressive load in a restrained column.

Due to this increase in the compressive load of a restrained column, the buckling temperature of the column is usually lower than that with unrestrained thermal expansion. The difference in temperatures at which a restrained and an unrestrained column buckles is dependent on the initial load in the column, its slenderness and the restraint stiffness. This temperature difference can be very high (i.e. several hundred degrees) and should be carefully considered. In general, the higher the restraint stiffness, the higher the additional compressive load and the lower the buckling temperature of a restrained column. A more slender column has a lower load carrying capacity and also attracts a higher additional compressive load due to a larger restrained thermal expansion. Both effects lead to a low buckling temperature for a restrained slender column.

However, when considering the behaviour of an axially restrained column, it is important to distinguish between the column buckling temperature and the column failure temperature. After buckling, the axial load in the column will reduce. Franssen (2000) suggests it is more appropriate to define the column failure temperature as the one at which the axial load in the column has returned to its original value. It is quite possible that what provides restraint to the column thermal expansion before buckling may also be able to provide the column with sufficient restraint to enable it to remain stable after buckling, albeit at a reduced axial load.

In order to quantify the failure temperature of a restrained column, it is important to include the post-buckling behaviour of the column.

5.3.2   Post-buckling behaviour of an axially restrained column

Figure 5.15 illustrates the complete load–temperature relationship of a restrained column, including the post-buckling behaviour. It may be divided into three stages: (i) the pre-buckling stage (A–B); (ii) the buckling stage (B–C); and (iii) the post-buckling stage (C–D). Equation (5.10) is only applicable to the pre-buckling stage A–B. For post-buckling analysis, an approximate method is presented here.

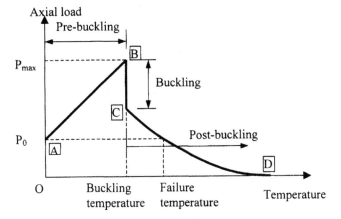

Figure 5.15 Complete load–temperature relationship of a restrained column in fire.

5.3.2.1   Load–temperature relationship at buckling (portion B–C in Figure 5.15)

When column buckling occurs, it may be assumed that the column under-goes a sudden lateral movement that is accompanied by an axial contraction as shown in Figure 5.16. The column temperature is fixed. If the column axial contraction is Δ_v, the total compressive load in the column becomes:

$$P = P_{max} - \frac{K_s K_c}{K_s + K_c} \Delta_v = P_{max} - K_s' \Delta_v \tag{5.11}$$

where P_{max} is the maximum compressive load in the column (cf. Figure 5.15) before buckling occurs.

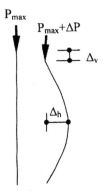

Figure 5.16 Forces and deflections at buckling.

The column axial contraction is due to the second order effect of column lateral deflection when buckling occurs. Generally:

$$\Delta_v = \phi \frac{\Delta_h^2}{L} \tag{5.12}$$

in which Δ_h is the maximum column lateral deflection at its mid-height. The column lateral deflection profile is approximately parabolic and $\phi = 8/3$.

Thus, the maximum bending moment in the column after buckling (point C in Figure 5.15) is:

$$M_{max} = P \times \Delta_h \tag{5.13}$$

After buckling and at point C in Figure 5.15, the column is stable and its axial load and bending moment satisfy the following interaction equation:

$$\frac{P}{P_{c,fi}} + \frac{M_{max}}{M_{p,fi}} = 1 \tag{5.14}$$

where $P_{c,fi}$ is the column buckling resistance and $M_{p,fi}$ is the plastic bending moment capacity of the column cross-section. Both $P_{c,fi}$ and $M_{p,fi}$ are evaluated at elevated temperatures.

Equation (5.14) is similar to the interaction equation in BS 5950 Part 1 (BSI 1990a) for checking the overall buckling of a column under combined axial load and bending moment where there is no LTB. Since the bending moment distribution in the column is not uniform, using the maximum bending moment in equation (5.14) will underestimate the column strength. The true influence of using the maximum bending moment will depend on the slenderness value of the column. Clearly, if the column is short and its flexural buckling resistance ($P_{c,fi}$) approaches its squash resistance, equation (5.14) is almost accurate. Of course, the approximate calculation method described here can be easily adapted when a reduced equivalent bending moment is used in equation (5.14).

Substituting equations (5.11–5.13) into equation (5.14), observing that $P_{max} = P_{c,fi}$ and that the solution for Δ_h should be non-trivial, the following quadratic equation can be solved to obtain the maximum lateral deflection of the column Δ_h:

$$\frac{\phi K_s'}{L \times M_{p,fi}} \Delta_h^2 + \frac{\phi K_s'}{L P_{c,fi}} \Delta_h - \frac{P_{max}}{M_{p,fi}} = 0 \tag{5.15}$$

Afterwards, the value of the maximum column lateral deflection Δ_h is substituted back into equations (5.12) and (5.11) to obtain the column axial load after buckling at point C in Figure 5.15.

5.3.2.2 Load–temperature relationship during the post-buckling stage (portion C–D in Figure 5.15)

During the post-buckling stage when the column temperature is increasing, the column axial deformation consists of three parts: the free thermal expansion, the increased mechanical deflection and the additional column contraction caused by the additional lateral deflection as shown in Figure 5.17. Thus, the additional compressive load in the column is obtained from:

$$\Delta P = \frac{K_s K_c}{K_s + K_c} (\Delta \varepsilon_{th} L - \Delta \varepsilon_{mec} L - \delta \Delta_v) \qquad (5.16)$$

where $\delta \Delta_v$ is the additional contraction of the column due to the additional lateral deflection of the column during the post-buckling stage.

For a small increase in the column lateral deflection $\delta \Delta_h$, equation (5.12) gives:

$$\delta \Delta_v = 2\phi \frac{\Delta_h}{L} \delta \Delta_h \qquad (5.17)$$

Substituting equation (5.17) into equation (5.16) gives the total axial load in the column as:

$$P = P' + \beta_1 - \beta_2 \delta \Delta_h \qquad (5.18)$$

where P' is the column axial load at the beginning of the temperature increment and

$$\beta_1 = \frac{K_s K_c}{K_s + K_c} (\Delta \varepsilon_{th} - \Delta_{mec}) L, \quad \beta_2 = 2\phi \frac{K_s K_c}{K_s + K_c} \frac{\Delta_h}{L} \qquad (5.19)$$

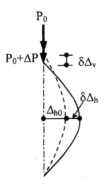

Figure 5.17 Post-buckling forces and deflections.

Substituting equation (5.19) into the column axial load–bending moment interaction equation (5.14), the following quadratic equation may be solved to give the increase in the column lateral deflection $\delta\Delta_h$:

$$\left(\frac{\beta_2}{M_{p,fi}}\right)\delta\Delta_h^2 + \left(\frac{\Delta_{h0}}{M_{p,fi}}\beta_2 - \frac{P' + \beta_1}{M_{p,fi}} + \frac{\beta_2}{P_{c,fi}}\right)\delta\Delta_h +$$

$$\left[1 - \frac{P' + \beta_1}{P_{c,fi}} - \frac{(P' + \beta_1)}{M_{p,fi}}\Delta_{h0}\right] = 0 \tag{5.20}$$

in which Δ_{h0} is the maximum lateral deflection of the column at the start of the temperature increment (see Figure 5.17). Afterwards, $\delta\Delta_h$ is substituted back into equations (5.17) and (5.16) to obtain the change in axial load ΔP of the column.

Because various terms in equation (5.20) are temperature dependent, the solution to this equation, thus the post-buckling load–temperature relationship of the column can only be obtained incrementally.

It should be noticed that during the buckling and post-buckling stages of the column behaviour, the column is contracting and unloading, therefore, the unloading stiffness of the restraint should be used.

Some results of the validation study for equations (5.11–5.20) are provided in Wang (2001b).

5.3.2.3   Some design implications

Using the method described above, failure temperatures of a restrained column have been calculated for various combinations of restraint stiffness, column slenderness and initial load level. Detailed results are described in Wang (2001b) so that only a summary of the conclusions are given here.

- For most practical applications where the restraint stiffness is low (2–3%) and the applied column load high (load ratio >0.5), there is little benefit in considering the post-buckling behaviour.
- However, with a combination of low initial load (load ratio <0.5), high column slenderness (slenderness >100) and high restraint stiffness (>5% of the initial column axial stiffness), the column failure temperature can be much higher than its buckling temperature. It is thus important to consider the column post-buckling behaviour in design.

Of course, if part of the initial load in the restrained column can be shed to other columns, the required load carrying capacity of the restrained column is reduced. The above approach can be directly used to evaluate the increased column failure temperature at this reduced load level.

5.3.3 Bending moments in restrained columns in fire

In addition to axial load, the bending moment of a restrained column can also change due to the variable stiffness of the heated and restrained column relative to that of the adjacent structure and to the P-δ effects. If the thermal expansion of the adjacent beam is restrained by the column, an axial force will develop in the beam and this force will also induce bending moments in the column (see Figure 5.12).

Consider the structural assemblies shown in Figure 5.18 where the thermal expansion of the adjacent structure is restrained. If the beam–column connections are ideally pinned, no bending moment will be transferred from the beam across the connection to the column. If the beam–column connections are rigid, the bending moment in the heated column will depend on the bending stiffness of the column relative to that of the adjacent beam. The relative stiffness of the column may change in a fire in a number of ways discussed below:

- If no local buckling occurs in either the restrained column or the adjacent beam (case A), the relative stiffness will essentially depend on the rates of heating in the column and the adjacent beam. In normal situations where the restrained column is connected to steel beams supporting concrete floor slabs, the slabs are usually at much lower temperatures. In addition, if the heated column is part of a continuous column that extends into the adjacent cold compartments, the cold column will have a much higher bending stiffness than the heated and restrained column in fire. Under these circumstances, the bending moment in the heated column will be considerably lower in the fire situation than at ambient temperature. On the other hand, in a portal frame type of structure where bare steel is used, if the column temperature is lower, the bending moment transferred to the column may be higher than the initial value.
- If restraint to the thermal expansion of the beam is high, a large compressive force can be generated in the beam and this can cause local buckling in the beam (case B). Since local buckling usually occurs at the beam ends, the connection to the restrained column becomes much more flexible and the bending moment transferred through the connection may be considerably reduced.
- If restraint to the thermal expansion of the column is high, a large compressive axial load may develop in the restrained column and cause local buckling in the column. Combined with the bending moment transferred from the beam, local buckling is likely to occur at the column top. This will make the column top act like a pin and release the column bending moment (case C).

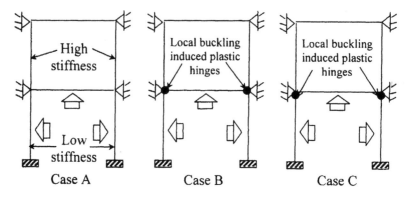

Figure 5.18 Variations of bending moment in a restrained column in fire.

Combining all the reasons above, it is most likely that in a continuous construction, bending moments transferred to a heated and restrained column will become considerably less in a fire such that they may be neglected in design calculations.

Now consider the effect of a heated column being pushed by lateral movement of the adjacent beam in a system as shown in Figure 5.19. The thermal expansion of the adjacent beam is restrained by the column to generate an axial load in the beam. This axial force can be evaluated using equation (5.10) and will induce bending moment distributions in the column

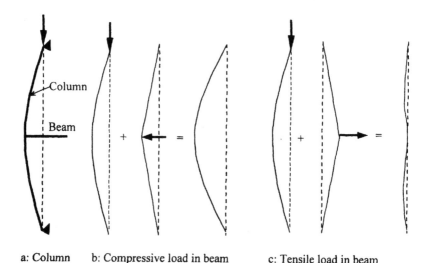

a: Column   b: Compressive load in beam   c: Tensile load in beam

Figure 5.19 Variations of bending moments in a column affected by the adjacent beam's thermal expansion.

as shown in Figures 5.19b and 5.19c. In addition to these bending moments, the axial force in the column will act on the lateral movement of the column and superimpose another set of bending moments as shown in these figures. Initially, these two sets of bending moments are additive and large bending moments can develop in the column (see Figure 5.19b). However, if the high axial load in the beam causes the beam to buckle, the resulting large lateral movement of the beam may make the axial force in the beam become tensile (Elghazouli and Izzuddin 2000; Usmani *et al.* 2001). Under this circumstance (Figure 5.19c), the bending moment in the column caused by the tensile force in the adjacent beam is opposing to the P-δ moments. Therefore, the level of net bending moment in the heated column will depend on the net effect of these two factors.

A limited study by Wang and Moore (1994) indicates that when the restrained column approaches failure and the axial force in the adjacent beam becomes tensile, the column failure temperature is only slightly lower than that without considering this effect. A more detailed theoretical study by Bailey (2000a) indicates that reductions in the column failure temperature can be much higher, up to 100 °C, depending on the structural configuration and the beam and column heating rates. A reduction of the column limiting temperature by 80 °C has been introduced in the recently published SCI/BRE design guide (Newman *et al.* 2000) for steel framed structures.

In studies of Wang and Moore (1994) and Bailey (2000a), the restraint effect of the floor slabs has not been included. In modern steel framed buildings with composite floor slabs, the composite floor slabs will exert a large restraint on the thermal expansion of the supporting beam such that the thermal expansion at the ends of the beam will be much lower than that occurring without the floor slabs. Consequently, the effect of the adjacent beam's thermal expansion on the column failure temperature is much lower than discussed in Section 5.33 based on bare steel frames. Since the effect on the failure temperature of a restrained column in a bare steel frame is moderate, it appears that in a more realistic structure with composite floor slabs, the effect of the adjacent beam's thermal expansion on the failure temperature of the connected column can be ignored.

In conclusion, the bending moment behaviour of a restrained column can be affected by the adjacent beams through bending moment transfer, P-δ effect and restrained thermal expansion in a complicated way. However, if the column failure temperature is the only design concern, bending moments in the column may simply be ignored.

5.3.4   Effective length of restrained columns in fire

At elevated temperatures, the stiffness of a column is low. If the adjacent structure remains cold and retains its original stiffness, the stiffness of the adjacent structure relative to that of the heated column can become very high so that a high degree of rotational restraint is provided to the heated column in a fire. This enhanced degree of rotational restraint reduces the column buckling length and increases the failure temperature of the column.

The structure adjacent to a heated column can either be the adjacent beams and floor slabs or the adjacent columns continuous from the column ends or both. In a simple construction, the beam to column connections are usually pinned and do not offer the heated column any restraint. With rigid beam–column connections, the adjacent beams and floors will give some rotational restraint to the column, but this restraint will depend on the rates of heating in the column and the adjacent beams. If the beams are at lower temperatures than the column, the rotational restraint of the adjacent beams and floors to the heated column will be higher than that at ambient temperature, otherwise it will be lower.

However, the most reliable rotational restraint to the heated column is from the adjacent cold columns outside the fire compartment. The procedure illustrated in Figure 5.20 may be used to evaluate the effective length of a heated column in a continuous construction.

- An appropriate structural assembly including the heated column is determined. Under the design loading and heating conditions, the failure temperature of the heated steel column can be calculated using advanced calculation methods such as those mentioned in Chapter 4.
- The column is then treated in isolation with simple supports at the column ends and the same loading condition as the heated column in the structural assembly. The column is analysed with different effective lengths and the different column failure temperatures are noted.
- The isolated column length that gives the same column failure temperature as that of the heated column in the structural assembly can then be taken as the effective length of the column in the structural assembly.

Using the above procedure, Wang *et al.* (1995), Wang (1997a) evaluated the effective length of a restrained steel column in non-sway and sway frames, for different combinations of beam length, column height, applied load and axis of bending. It was found that for non-sway frames, even though the effective length ratio of the column at ambient temperature ranged from 0.665 to 0.8, it changed within a narrow range of 0.5–0.6 under fire conditions, with a typical value of about 0.55. For sway frames, the column effective length ratio at ambient temperature varied from 1.3 to 2.0 and the value in fire situations changed between 1.0 and 1.2, with a typical value of 1.1. Thus this study concludes that under fire conditions, the rotational support of a heated column in a continuous structure approaches total fixity at ends.

The analysis described above has also been applied to evaluate the effective length of concrete filled columns in continuous construction (Wang

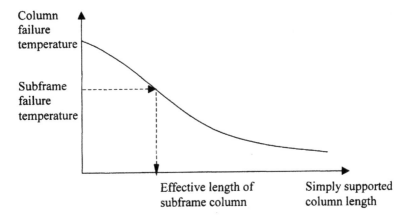

Figure 5.20 Procedure for determining the effective length of a rotationally restrained column.

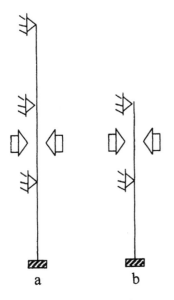

Figure 5.21 Two cases of continuous column.

1997c, 1999). This study ignored any beneficial restraining effects of the adjacent beams and floors and only considered the restraining effects of the adjacent cold columns. Therefore, this study used the structural assemblies shown in Figures 5.21a and 5.21b for columns with two-end and one-end continuous, respectively. Some results are shown in Figures 5.22 and 5.23. If the composite column was unreinforced, it was found that the

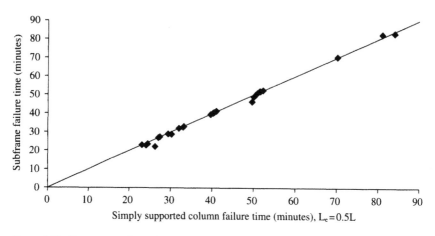

Figure 5.22 Comparison for columns continuous at both ends, unreinforced (from Wang 1999).

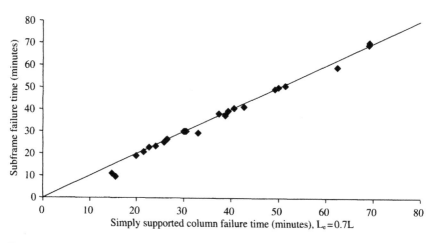

Figure 5.23 Comparison for columns continuous at one end, unreinforced (from Wang 1999).

effective length ratio of the column with two-ends continuous outside the fire compartment was close to 0.5. If the column was continuous outside the fire compartment at one end only, the effective length ratio was close to 0.7. These results have confirmed the conclusions of the study on steel columns. For a reinforced composite column, due to the relatively high stiffness of the reinforcement whose temperatures were relatively low, the column effective length was slightly lower than that obtained assuming total fixity at the column ends.

A similar study has also been carried out by Bailey (2000b) on concrete filled columns. The results of this study indicate that it is necessary to consider the problem of local buckling of the steel plate in concrete filled columns. Local buckling is a result of the steel tube resisting the applied load during the early stage of fire exposure (cf. Section 3.5.1). If the effect of local buckling is not considered, Bailey's results are similar to those of Wang described above. However, if local buckling is included, the rotational restraint to the column by the adjacent structure is reduced and the column effective length will be higher than those with fix-ended supports. The results of Bailey's study indicate that effective length ratios of 0.75 and 0.8 are more appropriate for a column with two-ends and one-end continuous from the fire compartment, respectively. Furthermore, for a column with one continuous end, this end should be at the bottom of the column. If the continuous end is at the top and local buckling occurs, the column effective length ratio should be taken as 1.0.

Bailey's study has raised an important point for which there is insufficient test information for reliable evaluation. He assumed that local buckling would occur at the top end of a restrained composite column so that the column top should be treated as a pin support as shown in Figure 5.24. Consequently, there is very little difference in the effective length ratio of a column with both ends continuous and with only the bottom end continuous. However, with a lack of detailed studies either experimental or theoretical, it is not certain where local buckling of the steel tube may occur. Test observations by Sakumoto *et al.* (1993) seem to suggest that local buckling of the steel tube would occur near the column center, rather than at the column end and the evidence of tests by Edwards (1998a) suggest that local buckling of the steel tube was almost random. Although local buckling

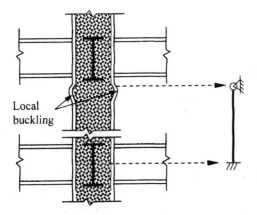

Figure 5.24 Conjectural effect of local buckling on the effective length of a concrete filled column.

has been observed in other fire tests (e.g. Lie and Chabot 1992), the locations in the column have often not been reported.

Nevertheless, considering the uncertainty about the location of local buckling in a concrete filled steel tube, and the theoretical observation from Bailey (2000b) that the column effective length may be adversely affected by local buckling of the steel tube at the column top, it is prudent that design considerations should err on the safe side.

5.4 SUMMARY OF COLUMN BEHAVIOUR IN FIRE

The behaviour of an isolated column in fire is relatively simple and is similar to that at ambient temperature. The effect of fire is simply to increase the axial deformation of the column. On the other hand, the behaviour of a restrained column in fire is complex. However, as far as FR design is concerned where the failure temperature of the column is the main interest, the following simple recommendations may be given:

1   The effect of axial restraint on the column axial load may be evaluated using equation (5.10);
2   In simple construction, bending moments from eccentricity should be considered. In continuous construction, bending moments may be ignored;
3   For a steel column and concrete encased composite column, the column end support may be assumed to be fixed if this end is continuous outside the fire compartment; and
4   For a concrete filled composite column, the bottom end may be assumed to be fixed if it is continuous outside the fire compartment and the top end should be assumed to be pinned.

5.5 BEHAVIOUR OF BEAMS IN FIRE

5.5.1 Statically determinate beams

When discussing the behaviour of a beam in fire, it is important to distinguish between different types of support of the beam. The behaviour of a simply supported beam usually forms the basis of any such discussion. Here the term "simple support" means that the beam is statically determinate and there is neither rotational nor longitudinal restraint to the beam.

Consider the flexural bending behaviour of a beam. It may be classified into two types: laterally restrained or laterally unrestrained. In a laterally unrestrained beam, if the beam is sufficiently long, bending of the beam about its major axis is accompanied by lateral movement in its weak axis and twist. This is called LTB and is similar to flexural buckling of a slender column. If LTB occurs, the bending moment resistance of the steel beam depends on the slenderness of the beam and is usually much lower than the plastic bending moment capacity of the cross-section of the beam. In a laterally restrained beam, the phenomenon of LTB does not occur and the bending resistance of the beam is governed by pure in-plane flexure. If local buckling does not occur, the plastic bending moment capacity of the cross-section can be reached.

LTB occurs in beams with open profile because the compression flange of the beam is not restrained. Fortunately, in realistic construction, it is often possible to identify restraint to the beam's compression flange so that there is no need to consider LTB. For example, in a simply supported steel beam supporting floor slabs, sufficient lateral restraint to the beam's compression flange is provided by the floor slabs. However, in some occasions, it may be difficult to apply lateral restraints to the compression flange of the beam and LTB has to be considered. Such cases include the hogging bending moment region of a continuous beam or the primary beams in a parallel beam system.

The behaviour of a laterally restrained beam and a laterally unrestrained beam in fire will be considered separately.

5.5.1.1   Laterally restrained beams

Since laterally restrained beams apply in a majority of cases of construction, numerous standard fire resistance tests have been carried out on them. A simple analysis of the results of one fire test can be used as an example to illustrate the behaviour.

Figure 5.25 shows the recorded time-vertical deflection of a typical beam (Wainman and Kirby 1987). Also shown in the figure are the calculated thermal bowing deflection in the beam; the difference between the measured total deflection and the calculated thermal bowing deflection (which may be regarded as the test mechanical deflection); and the calculated mechanical deflection. Assuming a constant coefficient of thermal expansion for steel and a linear distribution of temperature in the depth of the beam cross-section, the maximum thermal bowing in a simply supported beam is obtained from:

$$\delta_{th} = \frac{\alpha \times \Delta T \times L^2}{8d} \tag{5.21}$$

where α is the coefficient of thermal expansion of steel (0.000014 per °C), ΔT the temperature difference (in °C) between the lower and upper flanges, L the beam span and d the depth between the centroids of the upper and lower flanges (Figure 5.26).

The mechanical deflection is calculated by first finding out the curvature of the beam. calculated as the difference in mechanical strains of the lower

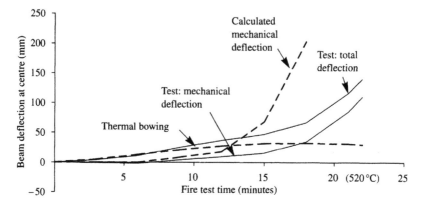

Figure 5.25 Deflection–time relationship of a simply supported beam in fire (from Wainman and Kirby 1987, Test 3). Reproduced with the permission of Corus.

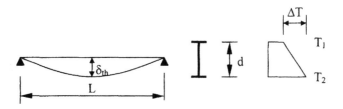

Figure 5.26 Thermal bowing in a simply supported beam.

and upper flanges divided by the beam depth. The upper and lower flange strains are obtained from the stress–strain relationships of steel at elevated temperatures (Kirby and Preston 1988) at a constant stress of $165\,N/mm^2$ applied at ambient temperature. This calculation procedure is shown in Figure 5.27. Clearly, as the temperature in the lower flange of the beam becomes much higher than the upper flange, the stress in the lower flange will be much lower than $165\,N/mm^2$ due to reduced lower flange stiffness and consequently, the lower flange strain will be much lower than that at a stress of $165\,N/mm^2$. This has not been taken into consideration, but the closeness between the calculated results using the simple calculation method and test results indicate that the beam behaviour can still be explained by assuming equal stresses in the lower and upper flanges of the beam.

It can be seen from Figure 5.25 that during the early stage of heating when the beam temperature is lower than $500\,°C$, the beam deflection is largely from thermal bowing caused by temperature gradient. The increase in the beam mechanical deflection is small. As the beam gets hotter but the temperature gradient stabilizes, the increase in beam deflection is mainly

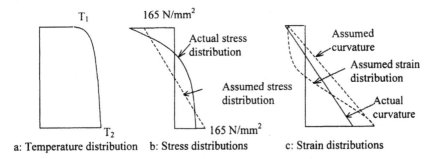

T_1 165 N/mm^2

Actual stress distribution

Assumed curvature

Assumed strain distribution

Assumed stress distribution

Actual curvature

T_2 165 N/mm^2

a: Temperature distribution   b: Stress distributions   c: Strain distributions

Figure 5.27 Procedure for calculating the mechanical deflection of a beam.

from reduced beam stiffness at high temperatures. At the end of the fire test, the beam temperature has become so high that the applied load cannot be supported by the reduced bending moment capacity of the beam. This is indicated in Figure 5.25 by the so-called "run-away" deflection of the beam.

It maybe noted that for a simply supported beam such as that shown in Figure 5.25, the run-away deflection occurs when the beam deflection is approximately span/20. This corresponds to one of the conditions to terminate the standard fire resistance test (Section 3.2.1). However, it should be pointed out that for other beam arrangements (as will be mentioned in later examples), beam run-away deflection may or may not occur and when it occurs, the deflection may be different from span/20.

Figure 5.25 shows that the unprotected steel beam has a standard fire resistance rating of about 20 min. This is typical of unprotected and isolated steel beams. Since the Building Regulations requirement for fire resistance is usually at least 30 min, and fire resistance tests are usually conducted on simply supported steel beams, it follows that almost all steel beams would need fire protection if the fire resistance of steel beams were to be assessed in this way.

Although the beam deflection behaviour depends on the temperature gradient in the cross-section, the beam's fire resistance, measured by the temperature of its lower flange at failure, is relatively insensitive to this temperature gradient (Burgess *et al.* 1990, 1991).

Since the fire resistance of an isolated but laterally restrained beam is governed by the bending moment capacity of the cross-section at elevated temperatures, the three measures discussed below may be considered to increase the beam's bending moment capacity so as to increase its fire resistance.

Using composite action

The behaviour of a composite beam is similar to that of a steel beam. The only difference is that when the applied load is the same, the composite beam will survive a more severe fire due to increased bending moment capacity of the composite cross-section.

Utilizing connection capacity

Simple connections are usually used in multi-storey steel framed buildings. These connections are designed on the assumption that they do not transfer any bending moment from the connected beam to the connected column so that the beam may be treated as simply supported. However, results of intensive research studies in the last three decades or so (Nethercot 1985) indicate that a nominally simple connection often has a substantial amount of bending moment capacity. Design rules have been developed to take into account the real load carrying capacity of simple connections using the so-called semi-rigid design method (Couchman 1997). However, the acceptance and use of this design method appears to be slow, perhaps due to the relative complexity of the design rules for what are perceived to be small real benefits, there already being high reserve of strength in design even on the basis of a simply supported beam.

The connection capacity will always exist whether the designer takes it into account or not. If this connection capacity is not used at ambient temperature, it gives an opportunity to use this "extra" strength to increase the fire resistance of the connected steel beam.

Figure 5.28 shows the bending moment diagram of a simply supported beam under a uniformly distributed load w in fire. To ensure stability of the beam in fire, the required bending moment capacity should exceed $wL^2/8$. Now assume that the nominally simple connection has a bending moment capacity of $M_{c,fi}$. According to the plastic design method, the required bending moment capacity of the beam is reduced to $wL^2/8-M_{c,fi}$. This reduced bending moment capacity may enable the steel beam to survive higher temperatures in a fire.

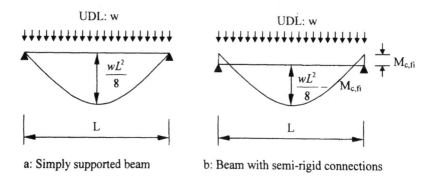

a: Simply supported beam   b: Beam with semi-rigid connections

Figure 5.28 Bending moments in a beam affected by semi-rigid connection.

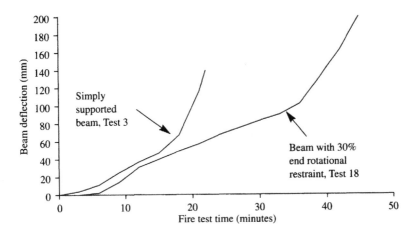

Figure 5.29 Effect of end rotational restrained on the behaviour of a steel beam (from Wainman and Kirby 1987). Reproduced with the permission of Corus.

For example, Figure 5.29 shows the time–deflection relationships of a steel beam with pin supports and with rotational restraints at the beam ends that sustain 30% of the free bending moment of the simply supported beam. The beam with rotational restraints at ends has a much higher fire resistance. From a commercial point of view, the simply supported beam needs fire protection to achieve a fire resistance rating of 30 min, whilst the rotationally restrained beam has an inherent fire resistance of 30 min without fire protection.

Theoretical studies have been carried out by a number of investigators to evaluate the possible benefits of using the bending moment capacity of nominally simple connections to enhance the fire resistance of the connected steel beam (Lawson 1990a; El-Rimawi *et al.* 1997; Liu 1998, 1999a). In these studies, the investigators have assumed that the connections are designed as simple joints at ambient temperature. They have found that by utilizing the available connection capacity in fire, the survival time of the connected steel beam is increased.

Whilst the connection bending moment capacity may be used to increase the fire resistance of a steel beam, the designer should bear in mind the following points:

- At present, there is very little information to enable reliable determination of the bending moment capacity of connections at elevated temperatures. As described in Section 3.8, the results of experiments on isolated connections cannot be used directly in realistic design due to the different behaviour of frame connections under fire conditions.

• The utilization of the connection bending moment capacity is on the assumption that a steel beam can transfer the connection bending moment at the beam ends. However, as described in Section 3.10.2.2, local buckling at the beam ends is likely to occur in realistic buildings in fire due to increased compressive load in the beam when its thermal expansion is restrained by the floor slabs. After local buckling, the bending moment that may be transferred at the beam ends will be much lower (Bailey 1998b; SCIF 1991). At present, when reliable information on the behaviour of frame connections in fire is not available, it is prudent not to use this possible source of reserve strength.

If the bending moment capacity of a connection is substantial and can be predicted reliably, the designer has a choice of whether to use the connection bending moment capacity at ambient temperature or under fire conditions. If the bending moment capacity of the connection is utilized at ambient temperature, a smaller steel beam may be used, alternatively higher loads could be applied. In any case, the steel beam would have a high load ratio. Only if there is an increase in the bending moment capacity of the connection relative to that of the steel beam in a fire over that at ambient temperature, were the effect of the connection beneficial. The results of a theoretical study by Liu (1999a) for flush end plate connections and extended end plate connections indicate that if the connection capacity is included in the ambient temperature design, no benefit can be gained from flush end plate connections and only small benefit in the case of extended end plate connections during subsequent fire conditions.

Using fire resistant steel

It is interesting to note in beams such as slim floor beams and partially encased beams a large parts of the steel section are at temperatures within the effective temperature range of FR steels. This makes it possible to employ FR steels to further increase the fire resistance of such beams. For example, Sha (1998) demonstrated that by using FR steels, slim floor beams could survive the standard fire exposure for 60 min even at the maximum loading. When the applied load in fire was relatively low, a standard fire resistance time of 90 min could be reached. When using FR steels, design calculations are no more complicated than those using conventional steels.

5.5.1.2 Behaviour of laterally unrestrained steel beams in fire

The problem of LTB of steel beams in fire has been studied by only a few authors. In particular, as indicated in Section 3.6.1, there is a lack of experimental results.

Unlike a laterally restrained beam whose failure is governed by the yield stress of steel, the behaviour and failure temperature of a laterally unrestrained beam are also affected by the stiffness of the beam. In particular, it is the tangent stiffness, rather than the initial Young's modulus, that quantifies the stiffness of a beam for LTB. If the applied stress on a beam is high, due to the non-linear stress–strain relationships of steel at high temperatures, the tangent stiffness of steel will be very low, leading to a low LTB temperature of the beam.

Bailey *et al.* (1996a) carried out a theoretical study to investigate the LTB temperatures of steel beams. In this study, the temperature distribution in the beams was assumed to be uniform. The effects of different parameters, such as bending moment distribution, location of loading, load ratio and beam slenderness were investigated. They found that steel beam failure temperatures for LTB could be much lower than the limiting temperatures given in BS 5950 Part 8 (BSI 1990b).

Vila Real and Franssen (2000) also reported the results of a theoretical study on LTB of steel beams in fire. In this study, they used the computer program SAFIR to calculate failure temperatures of unrestrained steel beams. A steel beam was subjected to heat exposure on all sides. Only the case of uniform bending moment distribution was considered in this study. The results of these numerical studies were then used to assess the predictions of Eurocode 3 Part 1.2 (CEN 2000b) for LTB. They found that the Eurocode approach did not give accurate results. They proposed an alternative approach based on the steel column design approach of Talamona *et al.* (1997) and found that the alternative approach gave better correlation with the computer results generated by SAFIR. Further details of this approach will be described in Section 8.3.4.

In summary, LTB of a steel beam in fire is an area where there is a lack of intensive study. Particularly, there appears to be no study of LTB of steel beams with non-uniform temperatures.

Summary of fire behaviour of statically determinate beams

The behaviour of a simply supported beam in fire is governed by pure bending and its deflection is comprised of two parts: (i) the mechanical deflection which increases with increasing temperatures due to reduced bending stiffness; and (ii) the thermal bowing deflection which depends on temperature gradients in the cross-section. The thermal bowing deflection is obtained from equation (5.21) and has very little effect on the behaviour of the beam during the later stage of heating. The failure of the beam is characterized by runaway deflection when the beam's bending stiffness has diminished to a negligible level. Provided that the loading condition is precisely defined, the fire resistance of a statically determinate beam may be calculated quite accurately by considering the plastic bending moment capacity of the cross-section for a laterally restrained beam or by using the LTB resistance for a laterally unrestrained beam.

5.5.2 Longitudinally and rotationally restrained beams in fire

When a beam forms part of a complete structure, the beam is restrained by the adjacent structure either longitudinally, or rotationally, or both. Because of various forms of structural interaction, the behaviour of such a beam is drastically different from that of an isolated beam without restraints. For such a beam, under the influences of high temperatures and temperature gradients, Figure 5.30 depicts the stages of behaviour that a restrained beam in fire may go through.

- When the rotational restraint acts on the temperature gradient of the beam, part of the thermal curvature of the beam will be restrained. The restrained part of thermal curvature will be converted into hogging bending moments at the beam ends (M_h, until stage 1) and the unrestrained part of thermal curvature will increase lateral deflections of the beam (δ_v, until stage 1).
- At same time, when the longitudinal restraint acts on the thermal expansion of the beam, the restrained part of thermal expansion of the beam will be converted into a compression force in the beam (P, until stage 1) and the unrestrained part of the beam's thermal expansion will increase the length of the beam (δ_h, until stage 1).
- When the hogging bending moment and the additional compression force in the beam at the ends are sufficiently high, compression stresses will be generated in the beam which can cause local buckling in the beam near the ends (stage 1).
- After local buckling, the compressive force in the beam will start to be relieved (P, region 1–2) accompanied by an increase in lateral deflection (δ_v, stage region 1–2) and a reduction in the length of the beam (δ_h, region 1–2). Due to the initial bending moments in the beam, this will be a gradual process, unlike that in a restrained column as described in Section 5.3.2 where there is a sudden increase in the column lateral deflection. The connection bending moments (M_h, stage 1–2) will also start to be relieved.
- When lateral deflections in the beam become sufficiently large (δ_v, after stage 2), the shortening in the beam length will overtake the beam's thermal expansion (δ_h, after stage 2) and thereafter, a tensile force (P, after stage 2) will develop in the beam and the beam is now under catenary action. In pure catenary action, the axial force in the beam will follow the yield stress of steel at increasing temperatures.

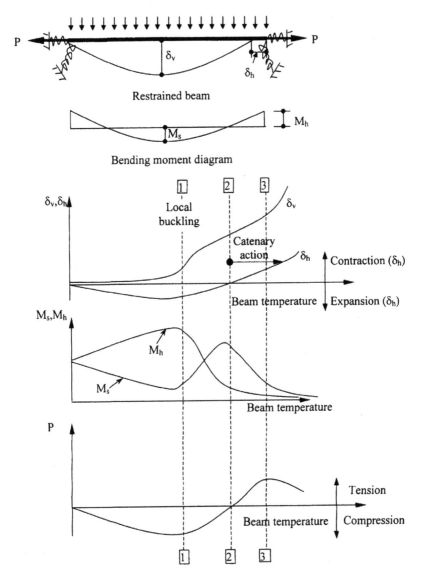

Figure 5.30 Illustrative behaviour of a rotationally and axially restrained beam in fire.

- The bending moment capacity of the beam will be negligible (M_h, M_s, after stage 3) but the beam will still be able to resist the applied load and will not experience "run-away" deflection as in a simply supported beam. Instead, the tensile force in the beam will be equal to the beam's decreasing tensile capacity at increasing temperatures (P, after stage 3) and the beam will resist the applied load by deflecting more (δ_v, after stage 3). At this stage, δ_v is almost completely accounted for by catenary sag.

- If very large deflections are acceptable, failure of the beam is likely to be governed by fracture of the connections to the adjacent structure.

Whilst in a simply supported beam lateral deflections of the beam are controlled by thermal gradient in the early stage and by increased mechanical deflection in the later stage of heating, this is no longer the case with a restrained beam. During the early stage of heating and before the beam goes into catenary action, the level of restraint (both axial and rotational) becomes the most important parameter and the effect of applied loading is relatively small (Ma and Makelainen 1999a,b; Elghazouli and Izzuddin 2000; Sanad *et al.* 2000a,b; Usmani *et al.* 2001).

From the above discussion, it is apparent that the behaviour of a restrained beam is highly complex due to complex non-linear geometrical, material and temperature interactions of the beam with the adjacent structure which exert both axial and rotational restraints. Usmani *et al.* (2001) correctly identified the importance of restraining the thermal strains and curvatures of a beam and presented an approximate analysis to illustrate the wide range of behaviour that may exist in a restrained beam. However, without using sophisticated finite element analysis, it has not been possible to find simple methods to quantify the behaviour of a restrained beam.

As a first order approximation, the complicated behaviour of a restrained beam may be by-passed if design is only concerned with the failure temperature of the beam. To do this, the restrained beam may be considered either in flexural bending or pure catenary action. If it is in pure catenary action, certain criteria will have to be checked and fulfilled so that the beam has sufficient capacity to develop catenary action, otherwise design should be limited to flexural bending only.

First consider the flexural behaviour of a restrained beam. In general, the beam is under combined axial load and bending moments and both hogging and sagging bending moments will be present in the beam. Moreover, the axial load and bending moments in the beam are variable and depend on the restraint condition at the beam ends and temperatures in the beam. However, as described above, the restrained beam is likely to experience local buckling at its ends. Consequently, it is likely that the axial load and the hogging bending moment will be relieved after local buckling with the beam becoming simply supported at the ends (SCIF 1991). Hence, the design method based on the behaviour of a simply supported beam may be used conservatively.

Now consider the behaviour of the restrained beam in pure catenary action where the beam has negligible bending resistance. The beam is in

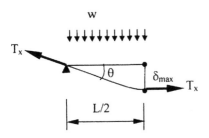

Figure 5.31 Catenary action in a restrained beam.

tension and the vertical component of the catenary force is in equilibrium with the applied vertical load on the beam. The deflection of the beam is determined on the basis of the beam's deflection profile and the applied vertical load. Thermal expansion of the beam and the level of restraint will not affect the beam deflection. This is in agreement with the results of a numerical analysis by Elghazouli and Izzuddin (2000).

5.5.3   Catenary action in a restrained beam in fire

Figure 5.31 shows half of a beam under a uniformly distributed load in pure catenary action. Assuming free rotations at the beam ends, the catenary force is related to the beam deflection by the equation:

$$T_x = \frac{wL^2}{8\delta_{max}} = \frac{wL}{8}\frac{L}{\delta_{max}} \qquad (5.22)$$

When the beam temperature is high, the catenary force available (T_x) is limited to the reduced tensile strength of the beam at the elevated temperatures. Therefore, with an increase in the beam's temperature and a reduction in the beam's strength, a higher beam deflection is required to resist the applied load wL. Equation (5.22) may then be used to estimate the maximum deflection δ_{max}. Alternatively, equation (5.22) may be combined with a limiting beam deflection to find the required catenary force, from which the beam failure temperature may be obtained.

5.5.3.1   Failure criteria

An isolated beam fails when it has reached its bending resistance, either by complete yielding of steel or by LTB, and run-away deflection occurs. However, in catenary action, the contribution of bending resistance is usually small and different failure criteria should be used. Failure of a restrained beam in catenary action may be considered to occur when any of the following happens:

- The beam deflection is too high so as to cause integrity failure of the building's FR compartmentation;
- The steel beam fractures;
- The adjacent structure (including connections) cannot resist the catenary force of the steel beam.

Since there is no reported observation of failure of a restrained beam in catenary action, conditions of beam failure cannot be precisely quantified at present. In the absence of further information, it appears to be reasonable to adopt the following suggestions:

- When deflections of the beam have to be limited to avoid integrity failure of the fire compartment construction, it is likely that the deflection limit is small and the behaviour and fire resistance of the restrained beam will be limited by the bending moment capacity of the beam in flexural bending. In this case, design should be based on a simply supported beam.
- Whether or not the steel beam will fracture will depend on the magnitude of mechanical strains. Fortunately, steel is ductile and its ductility is further enhanced at high temperatures. It is very difficult to envisage the fracture of a steel beam before other failure conditions occur first.
- The most important criterion will be whether or not the adjacent structure has sufficient load carrying capacity to resist the catenary force in the beam. To check this condition, the following situations may be considered:

    i   If the beam is tied to a strong structure such as the case of an interior beam in a continuous construction or a beam connected to bracing systems (Figure 5.32), the tying resistance of the connections should be checked.

    ii   If the beam is connected to columns (Figure 5.33), the resistance of the connected columns to the beam's catenary force should govern. If the columns do not have sufficient bending strength, they will have to go into catenary action to support the catenary force in the beam. However, this will depend on the columns being tied to vertically strong systems at the ends. Since the top of a column is usually free to move vertically, catenary action cannot realistically develop in columns. Thus, the only feasible way is to design columns for higher bending moment capacity (see Figure 5.33b).

    iii If the beam in catenary action is connected to other beams at right angles in plan (e.g. a secondary beam to primary beams as shown in

Figure 5.34, the connections are usually through the weak axis of these supporting beams. The supporting beams will usually not have sufficient bending resistance to resist the catenary force in the supported beam and will have to go into catenary action in plan. It is necessary to check whether this is possible.

Having speculated how a restrained steel beam may behave and fail in catenary action, a number of problems have to be addressed to make it possible to utilize this load carrying mechanism. The following lists the main assumptions and some of the issues that should be resolved by conducting more research studies.

1   Is catenary action a viable load carrying mechanism under fire conditions? The answer is almost certainly yes. The behaviour of some unprotected roof purlins in fire shown in Figure 5.35 clearly demonstrates the feasibility of this type of load carrying mechanism.

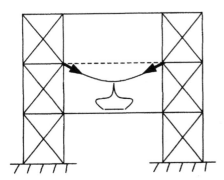

Figure 5.32 Beam in catenary action supported by strong bracings.

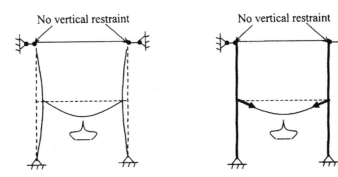

a: Beam connected to weak columns   b: Beam connected to strong columns

Figure 5.33 Catenary action in a beam connected to columns.

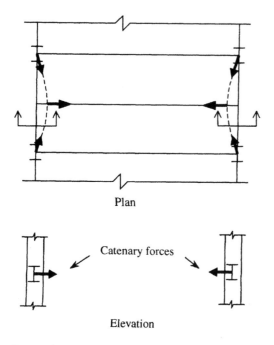

Plan

Catenary forces

Elevation

Figure 5.34 Catenary action in a beam connected to beams.

Figure 5.35 Roof purlins in catenary action (from Kirby et *al.* 1986). Reproduced with the permission of Corus.

2   Where will catenary action be applicable? This will depend on the support conditions of the adjacent construction as discussed above.

3   Will flexural bending action have influences on catenary action? The answer is probably yes. If the beam is securely anchored at the ends, catenary action will always occur. However, flexural bending will affect the magnitude of the beam's lateral deflection and the catenary force. At present, how flexural bending will affect catenary action in a restrained beam awaits answers from further extensive research studies still to be carried out.

4   Will the floor slabs be stable? The answer is yes because the slabs will still be supported by the steel beams that have gone into catenary action, even though their positions will have been displaced from being horizontal (shown as vertical in Figure 5.36a) to the deformed position as shown in Figure 5.36b.

Summary of restrained beams in fire

The behaviour of a restrained beam in fire is extremely complex and research is only starting to obtain a better understanding of the controlling factors that influence the beam behaviour. The uncertain transition area from flexural bending at small deflections to pure catenary action at very large deflections will almost certainly be the most difficult aspect of beam behaviour to be evaluated. In particular, the maximum catenary force is likely to be found during this stage. It is this maximum catenary force that determines whether or not the adjacent structure can allow catenary action to develop in the restrained beam. At present, sophisticated finite element programs have to be used and it is still too early to develop simplified design methods that are able to capture the main features of the complex beam behaviour.

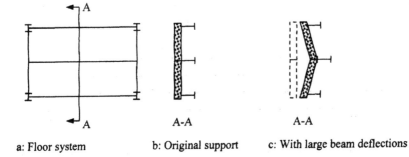

a: Floor system   b: Original support   c: With large beam deflections

Figure 5.36 Slab support at large beam deflections.

5.6 BEHAVIOUR OF SLABS IN FIRE

Depending on the boundary condition and dimensions of a slab, the applied load on the slab may be distributed from the slab to the surrounding structure in either one-way action or two-way action. If the slab aspect ratio exceeds about three, the effect of two-way action is likely to be small and the slab may conservatively be treated as spanning in one-way.

The behaviour of a one-way spanning slab is similar to that of a beam. Because there is no problem of local buckling or LTB, the bending resistance of a one-way spanning slab is governed by the formation of a plastic hinge mechanism. At large deflections, catenary action may develop in the slab, however, except for interior slabs, sufficient anchorage is unlikely to be available to allow catenary action to fully develop.

This section will focus on the behaviour of two-way spanning slabs, because this is the normal mode of slab behaviour in a building. Analogous to the behaviour of a beam, the behaviour of a slab at small deflections is flexural bending. At large deflections, membrane action, in particular, tensile membrane action, may develop in the slab. Ever since tensile membrane action was first identified and proposed to explain the superior performance of the large-scale fire tests at Cardington (cf. Section 3.10.2.2), the large deflection behaviour of composite slabs in multi-storey steel framed buildings has been the main focus of recent investigations. This section will summarize developments in this area and point out a few directions for possible future research.

5.6.1 Flexural bending behaviour of slabs at small deflections – yield line analysis

Yield line analysis (Johannsen 1962) has been widely adopted in the design of reinforced concrete slabs at ambient temperature. This analysis is analogous to plastic analysis of a framed structure. In this analysis, the most important task is to identify the yield line mechanism that requires the largest resistance moment. A yield line in a slab describes the condition when the slab cross-section along this line has reached its bending moment capacity.

For a simply supported isotropic rectangular slab under uniform loading, the yield line mechanism is shown in Figure 5.37. By using the principle of virtual work, the lower bound solution to the slab load carrying capacity may be obtained using equation (5.23) (Wood 1961).

$$M = \frac{wb^2}{24}\left[\sqrt{3 + \left(\frac{b}{a}\right)^2} - \frac{b}{a}\right]^2 \tag{5.23}$$

where M is the bending moment capacity of the slab per unit width; a and b are lengths of the longer and shorter spans of the slab; w is the uniform load density on the slab. For design purposes, Table 5.5 provides tabulated results of equation (5.23).

5.6.2   Membrane action in slabs at large deflections

5.6.2.1   Flexural bending strength of the slab in the Cardington test building

What had prompted recent research studies into membrane action in slabs was the remarkable performance of the Cardington test structure during severe fire attacks. As shown by the example below, even though the yield-line solution represents an upper bound estimation of the bending resistance of a slab, the applied load on the Cardington test structure was much higher than this upper bound solution.

Consider the BRE corner fire test (corner test 2 in Section 3.10.2.1). The bending resistance from yield line analysis is calculated as follows:

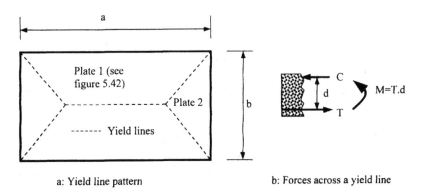

a: Yield line pattern   b: Forces across a yield line

Figure 5.37 Yield line analysis for a simply supported reinforced concrete slab.

Table 5.5 Yield-line solution for a simply supported slab

a/b	1.0	1.2	1.4	1.6	1.8	2.0	2.5	3.0	5.0	10	∞
wb^2/M	24	20.3	17.9	16.2	15.0	14.1	12.6	11.7	10.1	9.0	8.0

Source: Wood (1961).

Contribution from a simply supported reinforced concrete slab

Reinforcement yield stress:	$600 \, N/mm^2$
(note: the reinforcement was at low temperatures less than $400 \, °C$ and full strength may be used)	
Reinforcement force (A142 mesh):	$600 \, N/mm^2 \times 142 \, mm^2/m = 85\,200 \, N/m$
Concrete bending stress (C35 grade):	$0.67 \times 35 = 23.45 \, N/mm^2$
Depth of concrete block in compression:	$85\,200/(23.45 \times 1000) = 3.6 \, mm$
Depth of concrete cover to reinforcement:	$50 \, mm$
Length of lever arm:	$50 - 3.6/2 = 48.2 \, mm$
Slab sagging capacity:	$85.2 \times 48.2/1000.0 = 5.11 \, kN.m/m$
Aspect ratio:	$a/b = 9 \, m/6 \, m = 1.5$
Equation (5.23) gives:	$w = 17.1 \times 5.11/(6 \times 6) = 1.95 \, kN/m^2$

Contribution from flexural strength of the composite beam

Steel temperature (see Figure 3.29)	$900 \, °C$
Reduced steel strength:	$300 \, N/mm^2 \times 0.08 = 24 \, N/mm^2$
Tensile force in steel beam (356×171 UB51):	$24 \times 5150 \, mm^2/1000 = 123.6 \, kN$
Effective width of concrete:	$9000/4.0 = 2250 \, mm$
Depth of concrete block in compression:	$123\,600/(2250 \times 23.45) = 2.34 \, mm$
Depth of beam:	$303.8 \, mm$
Length of level arm:	$303.8/2.0 + 130 - 2.34/2.0 = 280.73 \, mm$
	$= 0.281 \, m$
Sagging moment capacity:	$0.281 \times 123.6 = 35.73 \, kN.m$
UDL on composite beam:	$8 \times 35.73/(9 \times 9) = 3.43 \, kN/m$
Equivalent UDL on slab:	$3.43/3.0 = 1.14 \, kN/m^2$

Therefore, the total flexural strength of the floor system including slabs and the composite beam is $1.95 + 1.14 = 3.09 \, kN/m^2$. This strength is much less than the applied load of $5.25 \, kN/m^2$ (see Section 3.10.2).

It is possible that other means, e.g. the slab hogging bending moment capacity and the steel decking, may have contributed somewhat to resisting the applied load, however, the total load carrying capacity in flexural bending would still not be sufficient. In any case, if the slab were nearing its collapse, an accelerating rate of beam/slab deflection would have been observed. The observed fact the slab showed no sign of imminent collapse (Figure 3.29) indicates that the applied floor loads were no longer sustained by flexural bending in the slab as has been assumed in the above calculations. Since the slab deflection was high, the only alternative explanation was that the slab was in membrane action.

5.6.3   Compressive membrane action

Of course, membrane action in slabs was not a new concept when it was used to explain the load carrying mechanism of the Cardington test structure.

Earlier studies of membrane action in reinforced concrete slabs go back about 50 years when Ockleston (1955) carried out loading tests on a defunct hospital building in Johannesburg and found that the strength of the floor slabs was much higher than that predicted by yield-line analysis. Ockleston (1958) identified the load carrying mechanism as arching action, or compressive membrane action.

Compressive membrane action relies on the vertical component of the compressive force in the slab to resist the applied load. As shown in Figure 5.38, if the centre of the compressive force at the support is at the lower corner of the slab (lowest position) and the centre of the compressive force in the slab is at the top surface (highest position), compressive membrane action may develop only if the slab vertical deflection does not exceed its depth. When the slab vertical deflection approaches its depth, the compressive force in the slab approaches the infinite in order for its vertical component to resist the downward vertically applied load. Once the slab vertical deflection exceeds the slab depth, the slab will lose its stability. Hence compressive membrane action has only a very limited range in which it is possible.

Obviously, the centres of compression forces in the slab will not be at the above assumed extreme positions. Under normal conditions, the allowable slab deflection is much lower before compressive membrane action becomes unstable. Usually, the allowable slab deflection is about 1/2 slab depth. Therefore, for compressive membrane action to be a viable load carrying mechanism, the slab has to be very deep. Moreover, in order to utilize compressive membrane action, strong external supports should be available to support the compressive forces in the slab. If these external supports are not available, such as the case of a simply supported slab without any in-plane resistance, compressive membrane action will not develop.

5.6.4   Tensile membrane action

In modern steel-framed structures with composite slabs, the structural floor slabs are usually thin and slab deflections (especially under fire conditions) are usually far greater than the slab thickness. This renders compressive

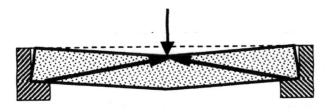

Figure 5.38 Compressive membrane action.

membrane action at best a transitional phenomenon in the slab behaviour. For such a slab to continue resisting increased loads, the only possibility is that the compressive membrane forces in the slab become tensile and the slab undergoes tensile membrane action.

When a slab is in tensile membrane action, the reinforcement in the slab centre is in tension and the slab resists the applied vertical loads by stretching of the reinforcement net under deflection. How the tension forces in the reinforcement net are anchored will depend on the in-plane restraint of the slab around the support. If the slab has sufficient external in-plane restraints around its edges, such as an interior slab in continuous construction, the slab tensile membrane forces will be resisted by the edge supports, as shown in Figure 5.39. In more realistic situations, the external supports at the slab edges will not be sufficiently strong to resist the tensile forces in the slab reinforcement net. Here tensile membrane action can still develop in the slab and the tensile forces in the reinforcement net are resisted by the slab forming a compressive ring beam around its edges. This load carrying mechanism is illustrated in Figure 5.40.

Therefore, depending on the external supports to the in-plane membrane forces of the slab, the complete load-deflection behaviour of the slab is as illustrated in Figure 5.41. Failure of the slab is usually indicated by fracture of the reinforcement.

It is clear from the above discussion that tensile membrane action can only occur when the slab undergoes very large deflections. This load carrying mechanism has not been used in the design of reinforced concrete slabs at ambient temperature, since only small deflections are acceptable in normal use. However, it has been used as a method to prevent progressive collapse

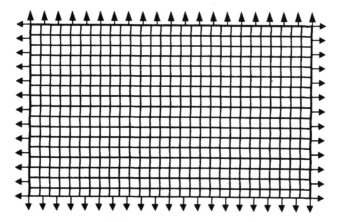

Figure 5.39 Mechanism of tensile membrane action in a plate with strong in-plane restraints at edges.

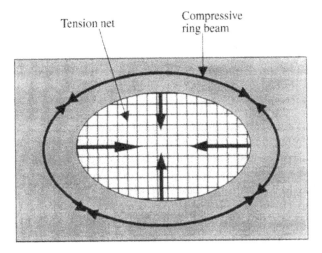

Figure 5.40 Mechanism of tensile membrane action in a plate without in-plane restraint.

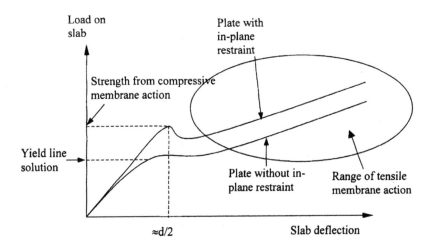

Figure 5.41 Complete load-deflection behaviour of a slab.

(Hawkins and Mitchell 1979; Mitchell and Cook 1983). Under fire conditions, the requirements for deflection control can be different. Because fire is an accidental condition, it is not necessary to limit the slab deflection, provided the large slab deflection does not breach the requirements of FR compartmentation so as to cause further fire spread. Thus tensile membrane action may be used under fire conditions. Indeed, most current research studies on slab behaviour in fire focus on how to predict reliably the slab load carrying capacity under tensile membrane action.

5.6.4.1 A summary of developments in tensile membrane action

A brief description of historical developments

Even though observations from the Cardington fire tests have sparked a renewed interest in tensile membrane action in reinforced concrete slabs, a literature review will reveal that this load carrying mechanism was investigated with some vigour and intensity in the 1960s. As a result, a number of analytical models have been provided. Notable developments are from Hayes (1968a,b) and Hayes and Taylor (1969), Kemp (1967), Park (1964a, b,c, 1965), Taylor (1965), Sawczuk and Winnicki (1965) and Wood (1961). Park's studies concentrated on slabs with in-plane restraints. He suggested that at collapse, the slab deflection could conservatively be limited to 1/10th of the slab span. Later research by Black (1973) found that the slab deflection limit could be increased to 0.15 times the slab span. Wood analysed a simply supported circular slab with isotropic reinforcement. Kemp extended the analysis of Wood to a simply supported square slab. Taylor provided a similar analysis to that of Kemp for a simply supported square slab, the difference being in the assumed positions of the slab neutral axis in their analyses. Sawczuk and Winnicki proposed a method for simply supported rectangular slabs with isotropic reinforcement. However, Hayes pointed out that Sawczuk's analysis implied boundary forces at edges of an unrestrained simply supported slab.

Hayes made the greatest contribution to the development of tensile membrane action theory. He critically examined limitations of other methods and proposed a general method for simply supported rectangular slabs with orthotropic reinforcement. Hayes also carried out a large series of slab tests. His comparisons between the measured and predicted results indicated that his method was very accurate. Hayes's method has recently been adopted by Bailey (2001) with minor modifications and developed into a design method under fire conditions (Bailey and Moore 2000a,b).

Hayes' analysis is based on the yield line pattern in Figure 5.37. Assuming rigid-plastic behaviour, only rigid body translations and rotations are allowed. With further assumptions that the neutral axes along the yield lines are straight lines and that the concrete stress-block is rectangular, variations of membrane forces along the yield lines become linear. Depending on whether or not the concrete has cracked through its depth in the central region, membrane force distributions are as shown in Figure 5.42.

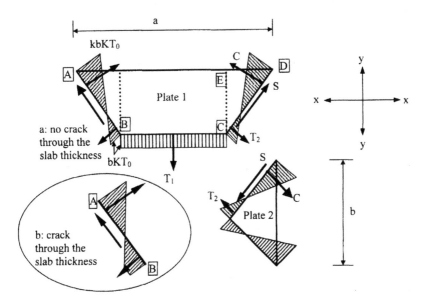

Figure 5.42 Membrane force distribution in a simply supported rectangular flat slab.

Detailed derivations of Hayes' theory are given in his Ph.D. thesis (Hayes 1968b). Only an outline of the procedure is given below.

1   The membrane stress distribution is shown in Figure 5.42a, where T_0 is the yield strength of reinforcement in the x direction and KT_0 that in the y direction. There are two unknown variables: (i) variable b depending on the amount of compressive force in concrete in the slab centre; and (ii) variable k to divide the slab into a compression zone near the slab edge and a tension zone near the slab centre. These two unknown variables can be solved by writing down two equilibrium equations: the equilibrium equation for membrane forces in the y-direction giving solution to variable k. The equilibrium in the x-direction is automatically satisfied. To find the solution to variable b, Hayes assumed that all reinforcements perpendicular to CE would completely yield and the slab was completely cracked through and took moment about point E using the membrane force distribution shown in Figure 5.43.

2   Having obtained values of b and k, the membrane force distribution in the slab along the yield lines are exactly defined. Hayes then applied the principle of virtual work to derive the slab load–deflection relationship, allowing for membrane action. Contributions to the internal virtual work consist of two parts: (i) those from the vertical components of the

membrane forces; and (ii) those from the flexural bending moments along the yield lines.

3   Relating the above solutions to the basic yield-line solution in flexural bending, Hayes derived enhancement factors, giving enhancements in the slab load carrying capacity due to membrane action. He derived the enhancement factors for the two plates shown in Figure 5.42. The enhancement factors for these two regions are in general not equal and Hayes adopted an averaging process for the overall enhancement factor for the slab.

4   Hayes used the same procedure to derive enhancement factors for a slab with a through-thickness crack and membrane force distribution as shown in Figure 5.42b.

A close examination of Hayes' equations will reveal that a through thickness crack will only develop in slabs of low aspect ratio (long span/short span less than about 1.3). Furthermore, the solutions for b and k are independent of the slab deflection, implying that the membrane stresses are constant. This is unlikely to be true and it is expected that as the slab deflects more, the extent of cracks that run through the slab thickness in the central region will be larger, which will force the concrete in the slab corners to develop higher compressive stresses. Indeed, concrete crushing in compression at the slab corners was used by Kemp (1967) as one of the slab failure criteria.

Hayes' loading tests were terminated due to large deflections and he did not comment on the failure condition of slabs under membrane action.

More recent developments

More recent developments on tensile membrane action come from Wang and Bailey. Wang (1996, 1997d) was the first to use tensile membrane action to explain the good performance of the Cardington fire test building and directly adopted the model of Kemp (1967) by assuming diagonal yield lines in a rectangular slab. Bailey *et al.* (2000) at the BRE carried out perhaps the first collapse test of a simply supported rectangular concrete slab. Based on

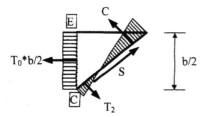

Figure 5.43 Membrane force distribution for bending moment about E.

the work of Hayes (1968b) and using the same procedure as outlined above, Bailey (2001) then developed a different model for tensile membrane action. Bailey's model is different from that of Hayes (1968b) in two aspects. First, he used the ultimate tensile strength instead of the yield strength of the reinforcement on the ground that slab failure was due to reinforcement fracture. Second, instead of assuming reinforcement failure along line CE in Figure 5.43, Bailey assumed this to happen at the slab centre based on observations of the BRE slab test. Bailey's model has now been incorporated in the new SCI design guide for multi-storey steel framed buildings in fire (Newman *et al.* 2000). In order to make this model applicable to design calculations, Bailey and Moore (2000a,b) introduced the following slab failure criteria:

$$\delta_{max} \le \frac{\alpha \Delta T b^2}{2.4 \times 8d} + \sqrt{\left(\frac{0.5 f_y}{E}\right)_{\text{reinforcement at } 20\,^\circ C} \times \frac{3b^2}{8}} \qquad (5.24)$$

or

$$\delta_{max} \le \frac{\alpha \Delta T b^2}{2.4 \times 8d} + \frac{b}{30} \qquad (5.25)$$

whichever is lower.

In equations (5.24) and (5.25), the first term represents the maximum thermal bowing deflection of the slab (cf. equation (5.21)) and a factor of 2.4 was obtained by curve-fitting available test data. The second term in equation (5.24) represents the maximum slab deflection to induce an average strain in the reinforcement to half its yield strain. The second term in equation (5.25) is to ensure that the allowable slab deflection does not exceed the Cardington fire test results.

The above mentioned models are analytical ones and can be easily used in hand calculations. They inevitably incorporate many simplifying assumptions. These simplifying assumptions may be removed if finite element analysis is employed. Work has recently started at the University of Sheffield using VULCAN, at Edinburgh University using ABAQUS and at Imperial College using ADAPTIC to investigate the behaviour of slabs at large deflections.

Main assumptions and further works

Analytical solutions will always be appealing because they offer an insight into the physical behaviour of the structure and their neat solutions lend them to adoption in practical design. Further researches are required to remove or improve the main simplifying assumptions so that designers have more confidence in their predictions. In particular, the following aspects deserve special attention:

1   *Sufficiency of vertical supports at the slab edges.* The implicit assumption in all analytical derivations is that the vertical supports at the slab edges do not deform. This is essential in order for the slab to deform in two-way double curvature. Whilst this is not a problem at ambient temperature, its implementation requires careful consideration under fire conditions. The vertical supports at the edges of a slab come from the edge beams. When these edge beams are under fire attack, they will deform. If their design is according to the current design methods (e.g. BSI 1990b or CEN 2000b), it is possible that these edge beams may experience large deflections which render the assumption of immovable vertical supports invalid. Further research studies are necessary to consider the effect of edge beam deflections on tensile membrane action in the slab and the maximum edge beam deflection that can be allowed while still achieving meaningful enhancement factors from tensile membrane action.

2   *Load sustained by the edge supports.* Under tensile membrane action, the stiffness of the slab is different from that in flexural bending. Therefore, the slab may not distribute its load to the supporting edge beams in the same way as at ambient temperature. In order to evaluate whether the supporting beams can provide sufficient vertical support to the slab, it is important that loads acting on the supporting beams are precisely evaluated.

3   *Failure criteria.* This is perhaps the most important assumption of analytical methods. The main output from the analytical solution is the slab load–deflection relationship. Not surprisingly, the slab failure criteria are expressed in terms of the maximum slab deflection. However, no slab collapse in fire has been observed to test the validity of these equations. Moreover, a particular problem is how to deal with local reinforcement fracture due to discontinuity of the slab at locations of through-thickness crack. The condition of the concrete in compression in the slab corners should also be included. Tensile membrane action can only develop if the slab deflections are very large, which may not be accommodated by other construction components. Slab failure criteria should therefore address the much wider issue of integrity failure of the building FR compartment which can lead to fire spread.

4   *Reliability of the compression ring beam.* Tensile membrane action in a simply supported slab develops in a self-contained manner in that the tensile forces in the reinforcement net are anchored by the compression ring beam. In addition to considering compressive crushing at the slab corners, the added complexity of thermal restraint by the adjacent structure should also be addressed.

The above solutions are based on a rectangular slab simply supported at the edges. Further work is required on other slab configurations, e.g. a slab supported on the edges and interiorly by a column. In addition, analytical solutions have necessarily been limited to simply supported slabs that are expected to have the lowest resistance. Whilst a slab with in-plane restraint around all edges has been dealt with by Park (1964a), other slabs with different types and degrees of restraint condition have not been considered so far.

5.7   OTHER ASPECTS OF FRAME BEHAVIOUR IN FIRE

Previous discussions have considered the behaviour of heated beams, columns, slabs and the immediately adjacent structure. Whilst this represents a huge advance over consideration of isolated beams and columns only, it is important to be aware the fire effects are more widespread. The behaviour of a complete building structure in fire is only starting to be investigated. Here only a qualitative description is given of a few aspects of the whole building behaviour in fire.

5.7.1   Remote areas of building

In fire safety design, the concern is usually for the heated part of the structure to survive fire attack. However, structural members far away from fire attack may also be affected. The plane frame test at Cardington (Section 3.10.2.2) provided a clear illustration of the possible extent of fire effect. The discussions in Sections 5.3 and 5.5.2 of this chapter have concentrated on the effects of restrained thermal expansion on the behaviour of heated columns and beams. The effects of these restraints can also be felt in remote parts of the building structure. Even though these remote parts of the building may still have sufficient load carrying capacity to keep the building structurally stable in fire, it is possible that extensive damage may be incurred in these remote parts of the structure.

5.7.2   Fire spread

In the analysis and design of a steel structure for fire safety, it is often assumed that fire is confined within the compartment of fire origin. The structural behaviour of the building is then studied under the compartment fire exposure. However, it is possible that fire may spread to other locations of the building. Bailey *et al.* (1996b) carried out a theoretical study and the results of their study indicate that the behaviour of a beam in the compartment of fire origin is affected by subsequent fire spread to adjacent fire compartments. Figure 5.44 may be used to explain the effect of fire spread.

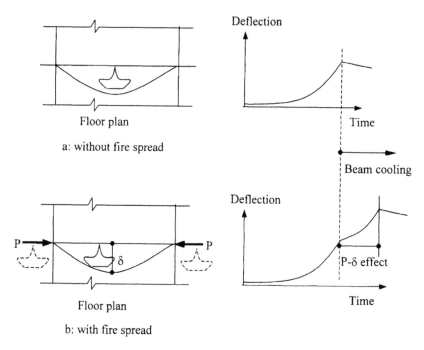

Figure 5.44 Effect of fire spread on beam deflection.

Without considering fire spread, the beam deflection reduces when it enters the cooling stage. However, if fire spread to the adjacent compartments is considered, the beam may undergo further deflections even though it is cooling down. This is due to the cooling beam being pushed by the hot beams in adjacent fire compartments which are under fresh fire attack after fire spread. After cooling, the beam can only recover the thermal bowing and elastic deformations. Thus the residual deflection in this beam will be greater if fire spread is considered. The immediate implication of this on design is the reparability of the beam. However, the more serious implication is the safety of the beam in the compartment of fire origin. Whilst the beam may survive the initial fire attack in the original fire compartment, subsequent pushing by the adjacent beams after fire spread may result in the failure of the beam during cooling in the original fire compartment. Liew et al. (1998) reported the results of a similar study and arrived at the same conclusions.

The same phenomenon may happen to a column in the compartment of fire origin when fire spread is vertical. If fire is spread to the compartment above, additional compressive load may be induced in the column in the original fire compartment even though it is cooling.

After the fire has spread to the adjacent fire compartments, whether or not structural members in the compartment of fire origin will remain safe during the cooling stage will depend on the interval between when cooling starts and when the subsequent fire reaches flashover in the adjacent compartments. If this time is relatively long, the initially affected structural members in the compartment of fire origin may have cooled sufficiently and recovered their ambient temperature strength. In this case, the danger of structural instability is very small. On the other hand, if this time interval is short as in the case of rapid fire spread, additional assessment of structural behaviour during the cooling phase is necessary.

5.7.3   Effect of bracing locations

In a simple construction, it is necessary to have bracing systems to provide lateral stability to the building structure. In the analysis of simple construction, it is often assumed that bracings provide ideal infinite support. In a theoretical study by Bailey *et al.* (1999a), they found that it was important to consider the stiffness and position of the bracing system when analysing the behaviour of a building structure. For example, Figure 5.45a and 5.45b show the results of their analyses for a cross-section of the Cardington building. In Figure 5.45a, infinite lateral supports are provided in the centre line of the building and in Figure 5.45b the realistic arrangement of cross bracing is considered. It is clear that with ideal infinite lateral support, the fire effect is concentrated within the heated fire compartment. With more realistic arrangement of bracing, the entire building is affected. The behaviour of the heated compartment is also different. With ideal infinite lateral support, the beam thermal expansion is concentrated on one side of the

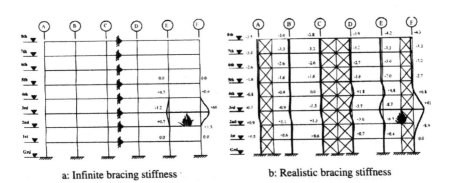

a: Infinite bracing stiffness   b: Realistic bracing stiffness

Figure 5.45 Effect of considering realistic behaviour of cross-bracing (Reprinted from the *Journal of Constructional Steel Research*, "The structural behaviour of steel columns during a compartment fire in a multi-storey braced steel-frame". Vol. **52**, Bailey *et al.* (1999) with permission from Elsevier Science).

compartment with very little movement on the side where the infinitely rigid lateral supports are located. With the more realistic arrangement of bracing, the flexibility of the bracing system attracts more displacement and the lateral displacement on the other side is reduced. This will have some influence on the bending moments in the heated columns from the P-δ effect.

5.7.4 Alternative load path

This chapter has focused on the behaviour of different structural elements, both when acting as isolated members or interacting with others. The load path is assumed to be clearly defined and the consideration has been to evaluate the time when the load path will become structurally unstable. However, it should be appreciated that in a realistic structure where there is a high degree of redundancy, there are many alternative load paths. The failure of one load path to transmit the applied load does not necessarily imply instability or collapse of the structure. The applied load on the structure may be resisted by another load path. If this alternative load path can be reliably identified, some structural members in the original load path may be allowed to "fail" without compromising safety of the structure. It is possible that alternative load paths can be incorporated in fire safety design in a more deliberate manner. Section 10.2.2 will provide an example of how to use alternative load paths.

5.8 CONCLUDING REMARKS

Spurred on by the large-scale structural fire tests at Cardington, major advances are starting to be made on the behaviour of steel structures under fire conditions. Two themes of recent research studies can be clearly identified:

1 It is essential to consider structural interactions in a highly redundant structure and these can be drastically different from those of statically determinate individual elements.
2 Large deflections can play a vital role in the survival of a steel framed structure in fire and the behaviour based on small deflections will not be sufficient to represent the whole picture.

Direct applications of the results from the Cardington fire tests are already having impact on the FR design of steel structures. However there is still some way to go before our understanding becomes sufficiently well developed to deliver comprehensive design guides. In particular, the author believes further researches in the following two areas would make significant contributions to our understanding of the behaviour of steel structures in fire. Both are related to large deflection behaviour:

1   *Catenary action of steel beams under fire conditions.* Future issues that need be resolved include the circumstances under which this load carrying mechanism will become possible, whether and how the various modes of bending failure preceding catenary action would affect catenary action and how designers may take advantage of this load carrying mechanism.

2   *Tensile membrane action in composite slabs.* Although the recently published new design guidance (Newman *et al.* 2000) makes use of this load carrying mechanism, a number of important issues still need to be resolved. They include reliability of the vertical supports, the effect of different slab configurations, reliability of the compression ring beam, effects of different boundary conditions and above all, the criteria of slab collapse.

Chapter 6

An introduction to heat transfer

This chapter will introduce heat transfer and its modelling in the context of structural engineering for fire safety. It will start from the simple topic of one-dimensional steady state heat conduction, leading to formulations of 3D transient state heat transfer equations to be used in various numerical heat transfer analyses.

There are three basic mechanisms of heat transfer: (1) conduction; (2) convection; and (3) radiation. In conduction, energy or heat is exchanged in solids on a molecular scale but without any movement of macroscopic portions of matter relative to one another. Convection refers to heat transfer at the interface between a fluid and a solid surface. Here the exchange of heat is due to fluid motion. This motion may be the result of an external force, causing the fluid to flow over the solid surface at speed. This is called forced convection. Convection can also occur due to buoyancy-induced flow when there is a temperature gradient in the fluid, causing a density gradient. This is called natural convection. Radiation is the exchange of energy by electromagnetic waves which, like visible light, can be absorbed, transmitted or reflected at a surface. Unlike conduction and convection, heat transfer by radiation does not require any intervening medium between the heat source and the receiver.

Except in the simplest cases, it is not possible to find analytic solutions to a heat transfer problem. Therefore, in fire safety applications, problems of heat transfer are usually solved either experimentally or numerically. Experiments are expensive to run and their results can only be applied to the specific situations that have been tested. Numerical analysis of heat transfer can be more general and is becoming more widely used. An understanding of the three heat transfer modes is essential for the selection of appropriate factors for carrying out numerical heat transfer analyses.

6.1   HEAT CONDUCTION

6.1.1   One-dimensional steady-state heat conduction

The basic equation for one-dimensional steady-state heat conduction is the Fourier's law of heat conduction. It is expressed as:

$$\dot{Q} = -k\frac{dT}{dx} \tag{6.1}$$

where, refering to Figure 6.1, dT is the temperature difference across an infinitesimal thickness dx. $\dot{Q}$ is the rate of heat transfer (heat flux) across the material thickness. The minus sign in equation (6.1) indicates that heat flows from the higher temperature side to the lower temperature side.

The constant of proportionality k is the thermal conductivity of the material. Values of thermal conductivity of steel, concrete and a number of generic fire protection materials will be given in Section 6.5. In many practical applications of fire safety engineering, the material thermal conductivity within the relevant temperature range may be approximated as a constant. Thus, equation (6.1) may be replaced by its finite difference equivalent:

$$\dot{Q} = -k\frac{T_2-T_1}{\Delta x}, \quad or, \quad T_1-T_2 = \dot{Q}\frac{\Delta x}{k} \tag{6.2}$$

where referring to Figure 6.2, T_1 and T_2 are temperatures at the two sides of a material and Δx is the material thickness. $\Delta x/k$ expresses the thermal resistance of the material.

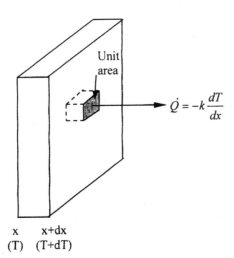

Unit area

$$\dot{Q} = -k\frac{dT}{dx}$$

x   x+dx
(T)   (T+dT)

Figure 6.1 Heat conduction in one dimension.

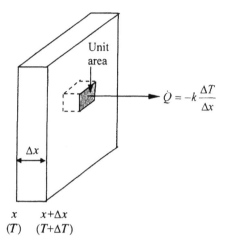

Figure 6.2 Temperature distribution with constant thermal conductivity.

6.1.2 One-dimensional steady-state heat conduction in a composite element

As an introduction to convective and radiant heat transfer to be presented in Sections 6.2 and 6.3, Figure 6.3 shows a construction element with a number of layers of different materials. Assume that the thermal conductivity of each material is temperature independent. According to the principle of energy conservation, i.e. the heat flow across each layer of material being the same, using equation (6.2), the following equation may be derived:

$$\dot{Q} = -k_{12}\frac{T_2 - T_1}{\Delta x_{12}} = -k_{23}\frac{T_3 - T_2}{\Delta x_{23}} = \cdots k_{i,i+1}\frac{T_{i+1} - T_i}{\Delta x_{i,i+1}}\cdots$$
$$= -k_{n,n+1}\frac{T_{n+1} - T_n}{\Delta x_{n,n+1}} \tag{6.3}$$

After rearrangement of equation (6.3), the following equations are obtained:

$$T_1 - T_2 = \dot{Q}\frac{\Delta x_{12}}{k_{12}} = \dot{Q}R_{12}$$

$$T_2 - T_3 = \dot{Q}\frac{\Delta x_{23}}{k_{23}} = \dot{Q}R_{23}$$

$$\cdots$$

$$T_i - T_{i+1} = \dot{Q}\frac{\Delta x_{i,i+1}}{k_{i,i+1}} = \dot{Q}R_{i,i+1} \tag{6.4}$$

$$\cdots$$

$$T_n - T_{n+1} = \dot{Q}\frac{\Delta x_{n,n+1}}{k_{n,n+1}} = \dot{Q}R_{n,n+1}$$

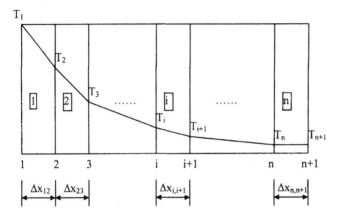

Figure 6.3 Steady-state conduction in a multi-layer composite element.

where for the ith layer: $k_{i,i+1}$ is the thermal conductivity; $R_{i,i+1}=\Delta x_{i,i+1}/k_{i,i+1}$ is the thermal resistance; T_i and T_{i+1} are temperatures on the two sides; and $\Delta x_{i,i+1}$ is the thickness.

Summing up equation (6.4) gives:

$$T_1 - T_{n,n+1} = \dot{Q} \sum_{i=1}^{n} R_{i,i+1} \tag{6.5}$$

Therefore, if temperatures on the two exterior sides of the composite (T_1 and T_{n+1}) are known, equation (6.5) may be used to obtain the heat flux Q. This value is then substituted back into equation (6.4) to obtain temperatures at the interfaces of different layers, i.e. $T_2, T_3, \ldots, T_n$.

6.1.3 Boundary conditions for one-dimensional heat conduction

In practical fire safety applications, exterior temperatures of a construction element are unknown variables. Instead, the surfaces of the element are in contact with fluids of known temperatures. For example, one side of the element may be in contact with the fire and the other side with the ambient temperature air. To determine the temperature distribution in the construction element, these fluid temperatures are used as boundary conditions.

In the example illustrated in Figure 6.4, when applying thermal boundary conditions, it is often assumed that the heat exchange between the fluid and

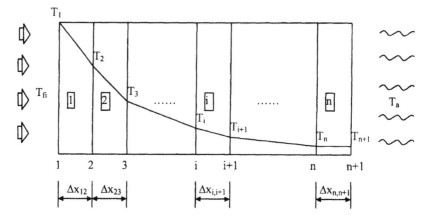

Figure 6.4 Boundary conditions for one-dimensional heat conduction.

the element surface is related to the temperature difference at the interface. Therefore on the fire side:

$$\dot{Q} = h_{\mathrm{fi}}(T_{\mathrm{fi}} - T_1) \tag{6.6}$$

on the ambient temperature air side

$$\dot{Q} = h_{\mathrm{a}}(T_{n+1} - T_{\mathrm{a}}) \tag{6.7}$$

where T_{fi} and T_{a} are the fire and air temperatures respectively.

Quantities h_{fi} and h_{a} are the overall heat exchange coefficients on the fire and air side respectively (see equation (6.57)), with values depending on convective and radiant heat transfer as discussed in the next two sections.

Equations (6.6) and (6.7) can be rewritten as:

$$(T_{\mathrm{fi}} - T_1) = \dot{Q}/h_{\mathrm{fi}} = \dot{Q}R_{\mathrm{f}}$$
$$(T_{n+1} - T_{\mathrm{a}}) = \dot{Q}/h_{\mathrm{a}} = \dot{Q}R_{\mathrm{a}} \tag{6.8}$$

where R_{f} and R_{a} may be regarded as the thermal resistance of the fire and ambient temperature air layers respectively.

Combining equations (6.8) and (6.5) gives:

$$T_{\mathrm{f}} - T_{\mathrm{a}} = R\dot{Q}, \quad \text{with} \quad R = R_{\mathrm{f}} + \sum_{i=1}^{n} R_{i,i+1} + R_{\mathrm{a}} \tag{6.9}$$

where R is the overall thermal resistance including the fire layer, the construction element and the ambient temperature air layer.

6.1.3.1   *Heat transfer coefficients*

It becomes clear that the main problem of heat transfer in fire is to determine appropriate heat transfer coefficients at the fluid/solid interface. A heat transfer coefficient contains two parts: (1) the convective part; and (2) the radiant part. Convective heat transfer applies only when the fluid is in contact with the solid surface. Radiant heat transfer will always occur whether or not the fluid is in contact with the solid surface.

6.2   CONVECTIVE HEAT TRANSFER

Heat convection is a difficult subject and its complete treatment is beyond the scope of this book. Instead, this section will only give, without derivation, the convective heat transfer coefficients that are necessary for modelling temperature distributions in structures in fire. More interested readers should consult thermodynamics textbooks such as Thomas (1980).

Types of flow   Fluid movement passing a solid surface is either forced convection or natural convection. Within each category, there are two types: the flow is either laminar or turbulent. If fluid movement is in a continuous path (streamline) without mixing with the adjacent paths, it is called laminar or streamline flow. On the other hand, if eddy motion of small fluid elements occurs, this produces fluctuations in the flow velocities of individual fluid elements, both in the direction of the surface and perpendicular to it, and introduces fluid mixing.

Dimensionless numbers   Studying heat transfer usually involves the use of a number of dimensionless numbers. These dimensionless numbers help extend the range of applicability of the limited number of small-scale experimental studies. The convective heat transfer coefficient (hereafter simplified as h_c) is related to the Nusselt (Nu) number using the following equation:

$$Nu = \frac{h_c L}{k}, \quad \text{giving} \quad h_c = \frac{Nu \times k}{L} \tag{6.10}$$

where L is the characteristic length of the solid surface and k the thermal conductivity of the fluid. The thermal conductivity of air at different temperatures is given in Table 6.1. Thus, the main task of convective heat transfer is to obtain the Nusselt number.

6.2.1   Heat transfer coefficients for forced convection

Table 6.2 gives relationships between the Nusselt number and other dimensionless numbers for forced convection.

Table 6.1 Property values of air at atmospheric pressure

$T(k)$	$\rho(kg/m^3)$	$C[kJ/(kg. °C)]$	$\mu \times 10^5$ [kg/(m.s)]	$\nu \times 10^6$ (m²/s)	K [W/(m °C)]	Pr
200	1.7684	1.0061	1.3289	7.490	0.01809	0.739
250	1.4128	1.0053	1.488	9.49	0.02227	0.722
300	1.1774	1.0057	1.846	15.68	0.02624	0.708
350	0.9980	1.0090	2.075	20.76	0.03003	0.697
400	0.8826	1.0140	2.286	25.90	0.03365	0.689
450	0.7833	1.0207	2.484	28.86	0.03707	0.683
500	0.7048	1.0295	2.671	37.90	0.04038	0.680
550	0.6423	1.0392	2.848	44.34	0.04360	0.680
600	0.5879	1.0551	3.018	51.34	0.04659	0.680
650	0.5430	1.0635	3.177	58.51	0.04953	0.682
700	0.5030	1.0752	3.332	66.25	0.05230	0.684
750	0.4709	1.0856	3.481	73.91	0.05509	0.686
800	0.4405	1.0978	3.625	82.29	0.05779	0.689
850	0.4149	1.1095	3.765	90.75	0.06028	0.692
900	0.3925	1.1212	3.899	99.3	0.06279	0.696
950	0.3716	1.1321	4.023	108.2	0.06525	0.699
1000	0.3524	1.1417	4.152	117.8	0.06752	0.702
1100	0.3204	1.1600	4.44	138.6	0.0732	0.704
1200	0.2947	1.179	4.69	159.1	0.0782	0.707
1300	0.2707	1.197	4.93	182.1	0.0837	0.705
1400	0.2515	1.214	5.17	205.5	0.0891	0.705
1500	0.2355	1.230	5.40	229.1	0.0946	0.705

Source: Thomas (1980).

Table 6.2 Convective heat transfer coefficients for forced convection

Flow type	Condition	Characteristic length	$Nu(=hL/k)$
Laminar flow, parallel to a flat plate of length L	$20 < Re < 3 \times 10^5$	L	$0.66Re^{1/2} Pr^{1/3}$
Turbulent flow, parallel to a flat plate of length L	$Re > 3 \times 10^5$	L	$0.037Re^{4/5} Pr^{1/3}$
Flow round a sphere of diameter L	General equation	L	$2 + 0.6Re^{1/2} Pr^{1/3}$

Source: Drysdale (1999), from An Introduction to Fire Dynamics, 2nd edition, Doughal Drysdale (1999). © John Wiley & Sons. Reproduced with permission.

In Table 6.2 Re and Pr are the Reynolds and Prandtl numbers respectively. The Reynolds number is defined by:

$$Re = \frac{\rho L U_0}{\mu} = \frac{L U_0}{\nu} \qquad (6.11)$$

where ρ is the fluid density, U_0 the flow velocity and μ the absolute viscosity of the fluid. Values of ρ and μ for air at different temperatures are given in Table 6.1. $\nu(=\mu/\rho)$ is the relative viscosity of the fluid and is also given in Table 6.1. The Reynolds number expresses the relative importance of the fluid momentum and the surface drag force due to viscosity. The faster the fluid velocity, the higher the Reynolds number and the higher the convective heat transfer.

The Prandtl number is defined as:

$$Pr = \frac{\mu C}{k} \tag{6.12}$$

where k is the thermal conductivity and C the specific heat of air, whose values are given in Table 6.1. The values of Pr are also given in Table 6.1. It can be seen that Pr is close to 0.7.

6.2.2   Heat transfer coefficients for natural convection

Under natural convection, heat exchange between the fluid and the solid surface depends not only on the fluid properties, but also on how the surface is placed in relation to the fluid, i.e. whether the surface is perpendicular or parallel to the fluid and whether the surface is above or below the fluid.

The general equation for the Nusselt number for natural convection is:

$$Nu = B \times Ra^m \tag{6.13}$$

where B and m are numbers whose values are given in Table 6.3.

Table 6.3 Convective heat transfer coefficients for natural convection

Surface configuration	Flow type	Condition	Characteristic dimension	B	m
Vertical plate (or cylinder) with height L,	Laminar	$10^4 < Ra < 10^9$	L	0.59	1/4
e.g. a wall	Turbulent	$Ra > 10^9$	L	0.13	1/3
Horizontal plate with area A and perimeter p,	Laminar	$10^5 < Ra < 10^7$	L for square	0.54	1/4
e.g. ceiling	Turbulent	$Ra > 10^7$	(L + W)/2 for rectangular	0.14	1/3
			0.9 D for circular disk		
			A/p for others		

Ra is the Raleigh number and is the product of the Grashof (Gr) number and the Prandtl number giving:

$$Ra = Gr \times Pr \tag{6.14}$$

The Grashof number is defined as:

$$Gr = \frac{gL^3 \beta \Delta T}{\nu^2} \tag{6.15}$$

where g is the gravity acceleration, β the coefficient of thermal expansion of the fluid and ΔT the temperature difference between the fluid and the solid surface.

According to the ideal gas law, the coefficient of thermal expansion of air at different temperatures is:

$$\beta = \frac{1}{T} \tag{6.16}$$

where T is the absolute temperature of air.

The Grashof number expresses the ratio of the upward buoyant force to the viscous drag. The higher the upward buoyancy, i.e. the higher the Grashof number, the faster the fluid movement and the higher the natural convective heat exchange coefficient.

6.2.3 Approximate values of convective heat transfer coefficients for fire safety

Evaluation of expressions in Tables 6.2 and 6.3 can become a lengthy process. In particular, since these expressions are temperature dependent and the surface temperatures are unknown variables in fire, an iterative process is required. Fortunately, in most cases of heat transfer analysis under fire conditions, radiation is the dominant mode of heat transfer and temperature calculations will not be very sensitive to even very large variations in the convective heat transfer coefficient. This renders it possible to use simplified methods.

6.2.3.1 Boundary condition on the fire side

Convective heat transfer in fire is mainly by natural convection. At the fire/solid interface, convective heat transfer is usually turbulent. Substituting values in Table 6.3 into equations (6.13) and (6.10) gives:

$$h_c = B\left[\left(\frac{g \times Pr}{T \times \nu^2}\right)\right]^{1/3} k(\Delta T)^{1/3} = \alpha(\Delta T)^{1/3} \tag{6.17}$$

Substituting values of B ($=0.14$), g ($=9.81\,\text{m/s}^2$), Pr, T, v and k into equation (6.17), the value of α may be found to vary between 1.0 to 0.6 within realistic fire temperatures of 400–1100 °C. For conservative calculations of temperatures of the fire exposed surface, the value of α may be taken as 1.0.

6.2.3.2 Boundary condition on the air side

Assuming laminar flow on the ambient temperature air side, a similar exercise gives:

$$h_c = \alpha(\Delta T)^{1/4} \qquad (6.18)$$

where α is approximately 2.2.

Eurocode 1 Part 1.2 (CEN 2000a) simplifies the calculations even further and recommends constant convective heat transfer coefficients as follows: on the fire side, $h_c = 25\,\text{W/m}^2$; on the air side, $h_c = 10\,\text{W/m}^2$.

6.3 RADIANT HEAT TRANSFER

When radiant thermal energy passes a medium, any object within the path can absorb, reflect and transmit the incident thermal radiation. Use absorptivity α, reflectivity ρ, and transmissivity τ to represent the fractions of incident thermal radiation that a body absorbs, reflects and transmits, respectively, giving:

$$\alpha + \rho + \tau = 1 \qquad (6.19)$$

6.3.1 Blackbody radiation

The three factors in equation (6.19) are functions of the temperature, the electromagnetic wave length and the surface properties of the incident body. Simplifications are usually made for fire engineering calculations. An extreme case is that all the incident thermal radiation is absorbed by the body i.e. $\alpha = 1$. Such an ideal body is called a blackbody.

6.3.1.1 Total power of thermal radiation

The blackbody thermal radiation is of fundamental importance to radiant heat transfer. It has many special properties, but the most important one being that it is a perfect emitter. This means that no other body can emit more thermal radiation per unit surface area than a blackbody at the same temperature. The total amount of thermal radiation (E_b) emitted by a blackbody surface is a function of its temperature only and is given by the Stefan–Boltzmann law:

$$E_b = \sigma T^4 \tag{6.20}$$

where σ is the Stefan–Boltzmann constant which is equal to 5.67×10^{-8} W/(m^{2}K^4) and T the absolute temperature in K.

6.3.1.2 Intensity of directional thermal radiation

Equation (6.20) gives the total thermal radiation of a unit area of blackbody surface (the emitter). This thermal radiation is not uniformly distributed in space. The directional dependence is expressed by the Lambert law:

$$I_\theta = I_n \cos \theta \tag{6.21}$$

In Figure 6.5, I_n is the intensity of thermal radiation in the normal direction to the emitting surface. I_θ is the intensity of thermal radiation in the direction at an angle θ to the normal direction of the emitting surface.

The intensity of thermal radiation is defined as the radiant heat flux per unit area of the emitting surface per unit subtended solid angle. Using this definition and equation (6.21), the intensity of directional thermal radiation can now be derived.

In Figure 6.6, the entire thermal radiation of a unit blackbody surface is covered by a hemispherical enclosure of radius r. Thus, the total thermal

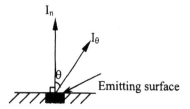

Figure 6.5 Directional intensity of radiant heat.

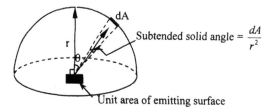

Figure 6.6 Determination of the intensity of thermal radiation.

radiation has to pass through the hemispherical enclosure, in other words, the total incident thermal radiation on the enclosure is equal to the total thermal radiation emitted by the blackbody surface.

For an infinitesimal surface area dA on the hemispherical enclosure, the subtended solid angle to the point of thermal radiation is dA/r^2. If the normal direction of this incident area makes an angle θ to the normal direction of the emitting blackbody surface, the incident thermal radiation on this area dA is:

$$dQ_\theta = I_\theta \frac{dA}{r^2} = I_n \frac{dA}{r^2} \cos\theta \qquad (6.22)$$

Integration of equation (6.22) over the entire hemisphere surface gives the total thermal radiation incident on the hemisphere. Equate this to the total thermal radiation of the unit area of emitting surface, giving:

$$E_b = \oiint dQ_\theta = \int_0^{\pi/2} I_n \cos\theta \frac{2\pi \times r \sin\theta \times rd\theta}{r^2} = \pi I_n \qquad (6.23)$$

Therefore, the intensity of thermal radiation in the normal direction to the blackbody emitting surface is:

$$I_n = \frac{E_b}{\pi} \qquad (6.24)$$

6.3.1.3 Exchange of thermal radiation between blackbody surfaces

Having obtained the directional intensity of thermal radiation, the radiant thermal exchange between two blackbody surfaces can now be calculated. Figure 6.7 shows two blackbody surfaces A_1 and dA_2, with A_1 being the

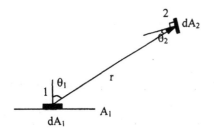

Figure 6.7 Radiant heat exchange between a finite and infinitesimal area.

emitting surface. It is required to find the incident thermal radiation on dA_2. In Figure 6.7, dA_1 is a small area on A_1 and points 1 and 2 are at the centres of dA_1 and dA_2. θ_1 is the angle between the normal to dA_1 and line 1–2 and θ_2 the angle between the normal to dA_2 and line 1–2.

If the total thermal radiation per unit surface area of A_1 is E_{b1}, the intensity of incident thermal radiation in direction 1–2 is $(E_{b1}/\pi)\cos\theta_1$. The effective area of dA_2 normal to direction 1–2 is $dA_2 \times \cos\theta_2$, giving a subtended solid angle of dA_2 on the centre of dA_1 of $dA_2 \times \cos\theta_2/r^2$. Therefore, the thermal radiation from dA_1 incident on dA_2 is:

$$d\dot{Q}_{dA_1 \rightarrow dA_2} = E_{b1}\frac{\cos\theta_1\cos\theta_2}{\pi r^2}dA_1 dA_2 \tag{6.25}$$

and the total thermal radiation from A_1 incident on dA_2 is:

$$\dot{Q}_{A_1 \rightarrow dA_2} = \int_{A_1} E_{b1}\frac{\cos\theta_1\cos\theta_2}{\pi r^2}dA_1 dA_2 = \Phi E_{b1} dA_2 \tag{6.26}$$

6.3.1.4   Configuration (view) factor Φ

On the right hand side of equation (6.26), $E_{b1}dA_2$ is the maximum incident thermal radiation on dA_2 and this occurs when dA_2 is completely surrounded by A_1. In most cases, the incident thermal radiation on dA_2 is much less. The factor Φ is used to represent the fraction of thermal radiation from A_1 incident on dA_2. This factor is often referred to as the configuration or view factor because it only depends on the spatial configuration between A_1 and dA_2. The configuration factor will not be greater than 1.0.

The configuration factor is an important value. However, the integration contained in equation (6.26) can be laborious to carry out. Since it depends only on the spatial arrangement of a surface (A_1) and a view point (dA_2), values of the configuration factor for many situations have already been calculated and presented in a number of textbooks. For example, interested readers may consult the SFPE handbook (SFPE 1995) for a variety of geometric arrangements. Table 6.4 gives the configuration factor from a rectangular plane to a view point (dA) opposite one of its corners as shown in Figure 6.8.

The configuration factor is additive, i.e. the configuration factor for a complex surface may be obtained by dividing the surface into a number of simple ones whose view factors can be obtained from readily available prepared tables. The total configuration factor is simply the summation of all the configuration factors of the sub-divided areas. For example, Table 6.4 may be used to find the configuration factor for a viewpoint opposite any point on a rectangular plane as shown in Figure 6.9. To do this, the

Table 6.4 Values of configuration factor for various values of α and S for situation in Figure 6.8 ($S = L_2/L_1$, $\alpha = |L_1 \times L_2|/D^2$)

α	$S=1$	$S=0.9$	$S=0.8$	$S=0.7$	$S=0.6$	$S=0.5$	$S=0.4$	$S=0.3$	$S=0.2$	$S=0.1$
2.0	0.178	0.178	0.177	0.175	0.172	0.167	0.161	0.149	0.132	0.102
1.0	0.139	0.138	0.137	0.136	0.133	0.129	0.123	0.113	0.099	0.075
0.9	0.132	0.132	0.131	0.130	0.127	0.123	0.117	0.108	0.094	0.074
0.8	0.125	0.125	0.124	0.122	0.120	0.116	0.111	0.102	0.089	0.067
0.7	0.117	0.116	0.116	0.115	0.112	0.109	0.104	0.096	0.083	0.063
0.6	0.107	0.107	0.106	0.105	0.103	0.100	0.096	0.088	0.077	0.058
0.5	0.097	0.096	0.096	0.095	0.093	0.090	0.086	0.080	0.070	0.053
0.4	0.084	0.083	0.083	0.082	0.081	0.079	0.075	0.070	0.062	0.048
0.3	0.069	0.068	0.068	0.068	0.067	0.065	0.063	0.059	0.052	0.040
0.2	0.051	0.051	0.050	0.050	0.049	0.048	0.047	0.045	0.040	0.032
0.1	0.028	0.028	0.028	0.028	0.028	0.028	0.027	0.026	0.024	0.021
0.09	0.026	0.026	0.026	0.026	0.025	0.025	0.025	0.024	0.022	0.019
0.08	0.023	0.023	0.023	0.023	0.023	0.023	0.022	0.022	0.020	0.017
0.07	0.021	0.021	0.021	0.021	0.020	0.020	0.020	0.019	0.018	0.016
0.06	0.018	0.018	0.018	0.018	0.018	0.017	0.017	0.017	0.016	0.014
0.05	0.015	0.015	0.015	0.015	0.015	0.015	0.015	0.014	0.014	0.013
0.04	0.012	0.012	0.012	0.012	0.012	0.012	0.012	0.012	0.011	0.010
0.03	0.009	0.009	0.009	0.009	0.009	0.009	0.009	0.009	0.009	0.008
0.02	0.006	0.006	0.006	0.006	0.006	0.006	0.006	0.006	0.006	0.006
0.01	0.003	0.003	0.003	0.003	0.003	0.003	0.003	0.003	0.003	0.003

Source: Drysdale (1999), from *An Introduction to Fire Dynamics*, 2nd edition, Doughal Drysdale (1999). © John Wiley & Sons. Reproduced with permission.

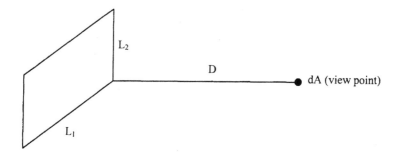

Figure 6.8 View point from a point opposite one corner of a parallel rectangular.

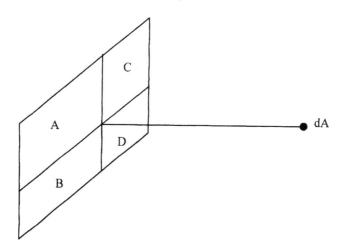

Figure 6.9 Illustration of the additive rule of configuration factor (equation (6.27)).

rectangle is divided into four smaller rectangles with the viewpoint lying perpendicular to one corner point of each of these smaller rectangles. The total configuration factor is:

$$\Phi = \Phi_A + \Phi_B + \Phi_C + \Phi_D \tag{6.27}$$

where Φ_A, Φ_B, Φ_C and Φ_D may be obtained using Table 6.4.

6.3.1.5   Exchange area

Equation (6.26) gives the incident thermal radiation from a finite area to a point. In realistic cases, it is often required to calculate the incident thermal radiation from one finite area to another finite area. For two finite areas A_1 and A_2, equation (6.26) now becomes:

$$\dot{Q}_{A_1 \to A_2} = \iint_{A_1,A_2} E_{b1} \frac{\cos \theta_1 \cos \theta_2}{\pi r^2} dA_1 dA_2 = F_{1 \to 2} A_1 E_1 \qquad (6.28)$$

The factor $F_{1 \to 2}$ is the "integrated configuration factor". It represents the fraction of radiant heat flux emitted by surface A_1 and incident on surface A_2. Since $A_1 E_1$ is the total thermal radiation from surface A_1, this is the maximum that can be incident on A_2. Therefore, $F_{1 \to 2}$ cannot be greater than 1. Similar to the configuration factor, values of the integrated configuration factor for many arrangements of two surfaces have been calculated and are available from textbooks, e.g. SFPE (1995).

6.3.2   Radiant heat transfer of greybody surfaces

No real material emits and absorbs radiation according to laws of the blackbody. In general, an additional term is necessary to define the radiant energy of an emitting surface. This is the emissivity ε. This term is defined as the ratio of the total energy emitted by a surface to that of a blackbody surface at the same temperature. Thus, the total radiant energy emitted by a general surface is:

$$E = \varepsilon \sigma T^4 \qquad (6.29)$$

In general, the emissivity of a surface depends on the wavelength of radiant energy, the temperature of the surface and the angle of radiation. However, if the emissivity is independent of these factors, the radiant surface is called a greybody surface. Greybody radiation is adopted in fire engineering calculations. According to Kirchhoff's law, the emissivity of a surface is equal to its absorptivity. Therefore, if there is no transmission of radiant energy through a greybody surface, the reflectivity of the surface is $1 - \varepsilon$, according to equation (6.19).

6.3.2.1   Heat exchange between two greybody surfaces

With the introduction of emissivity and reflectivity, the heat exchange between two greybody surfaces is complicated and analytical equations can be obtained only for very simple cases. One such a case is that of two very large parallel plates of area A, whose distance apart is small compared with the size of the plates so that radiation at their edges is negligible. Another case is that of two bodies with one inside the other which are a small distance apart compared with their dimensions, so that their areas are effectively equal. The latter case may be regarded to represent the situation of a fire

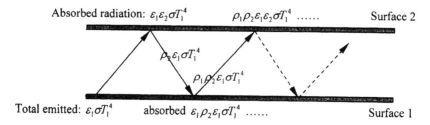

Absorbed radiation: $\varepsilon_1\varepsilon_2\sigma T_1^4$ $\rho_1\rho_2\varepsilon_1\varepsilon_2\sigma T_1^4$ ......   Surface 2

$\rho_2\varepsilon_1\sigma T_1^4$

$\rho_1\rho_2\varepsilon_1\sigma T_1^4$

Total emitted: $\varepsilon_1\sigma T_1^4$ absorbed $\varepsilon_1\rho_2\varepsilon_1\sigma T_1^4$ ......   Surface 1

Figure 6.10 Radiant heat transfer between two parallel infinite greybody surfaces.

surrounding a construction element. The analytical solution of these two cases may be obtained by referring to Figure 6.10.

Consider heat emitted by surface 1 and incident on surface 2. The radiant heat emitted by the greybody surface 1 at a temperature of T_1 is $\varepsilon_1\sigma T_1^4 A$, A being the surface area. Of this $\alpha_2\varepsilon_1\sigma T_1^4 A = \varepsilon_2\varepsilon_1\sigma T_1^4 A$ (since $\varepsilon_2 = \alpha_2$) is absorbed by surface 2 and $\rho_2\varepsilon_1\sigma T_1^4 A$ is reflected back to surface 1. Of this reflected radiation, surface 1 absorbs $\varepsilon_1\rho_2\varepsilon_1\sigma T_1^4 A$ and reflects $\rho_1\rho_2\varepsilon_1\sigma T_1^4 A$. From the initial radiation emitted by surface 1, the absorbed radiation on surface 2 is $\varepsilon_1\varepsilon_2\sigma T_1^4 A$. The reflected radiation starts the second round and the next absorbed radiation on surface 2 is $\rho_1\rho_2\varepsilon_1\varepsilon_2\sigma T_1^4 A$. This process of radiation (surface 1) – absorption and reflection (surface 2) – absorption and reflection (surface 1) goes on indefinitely. The total incident radiation on surface 2 is given by the following infinite series:

$$\dot{Q}_{1\to2} = \varepsilon_1\varepsilon_2\sigma T_1^4 A\left\{1 + \rho_1\rho_2 + (\rho_1\rho_2)^2 + \cdots\right\} \quad (6.30)$$

Similarly, the incident radiation on surface 1 from the emitted radiation by surface 2 is:

$$\dot{Q}_{2\to1} = \varepsilon_1\varepsilon_2\sigma T_2^4 A\left\{1 + \rho_1\rho_2 + (\rho_1\rho_2)^2 + \cdots\right\} \quad (6.31)$$

The energy exchange between surfaces 1 and 2 is therefore:

$$\dot{Q} = \dot{Q}_{1\to2} - \dot{Q}_{2\to1} = \varepsilon_1\varepsilon_2\sigma(T_1^4 - T_2^4)A\left\{1 + \rho_1\rho_2 + (\rho_1\rho_2)^2 + \cdots\right\} \quad (6.32)$$

Since $\rho_1\rho_2 < 1$, the series in the curly brackets of equation (6.32) is converging and its sum is $1/(1-\rho_1\rho_2)$. The radiant energy exchange is therefore:

$$\dot{Q} = \frac{\varepsilon_1\varepsilon_2}{1 - \rho_1\rho_2}\sigma(T_1^4 - T_2^4)A \quad (6.33)$$

Assuming τ to be negligible whence from equation (6.19), $\rho = 1 - \alpha$ (and by Kirchhoff's law $\rho = 1 - \varepsilon$), after some mathematical manipulation, equation (6.33) becomes:

$$\dot{Q} = \frac{1}{1/\varepsilon_1 + 1/\varepsilon_2 - 1} \sigma(T_1^4 - T_2^4)A = \varepsilon_r \sigma(T_1^4 - T_2^4)A \qquad (6.34)$$

In equation (6.34), ε_r is often referred to as the resultant emissivity and

$$\varepsilon_r = \frac{1}{1/\varepsilon_1 + 1/\varepsilon_2 - 1} = \frac{\varepsilon_1\varepsilon_2}{\varepsilon_1 + \varepsilon_2 - \varepsilon_1\varepsilon_2} \qquad (6.35)$$

For fire safety applications, it is often assumed that combustion fire gases and construction elements are in close contact so that they can be treated as two infinitely large surfaces and equation (6.34) may be applied.

Eurocode 1 Part 1.2 (CEN 2000a) gives the emissivity of fire and a general construction element surface as 0.8 and 0.7. Approximately, the resultant emissivity is calculated using:

$$\varepsilon_{res} = \gamma_{n,r}\varepsilon_{fi}\varepsilon_m \qquad (6.36)$$

where ε_{fi} is the emissivity of fire and ε_m the emissivity of the construction material. $\gamma_{n,r}$ is a modification factor for different standard fire testing furnaces. According to Kay et al. (1996), $\gamma_{n,r}$ is 0.45 for steelwork in the standard fire. For steelwork in natural fires, $\gamma_{n,r}$ is 1.0.

The difference in the resultant emissivity between the standard fire exposure and natural fires is attributed to the difference in emissivity of the combustion gas in the standard fire exposure and in natural fires. In a standard fire test, the combustibles are well mixed and burning is efficient so that the flame emissivity is low. In natural fires, there is a large amount of smoke and soot so that emissivity of the combustion gas is high.

Using equation (6.6), the radiant heat transfer coefficient is obtained as:

$$h_r = \varepsilon_r \sigma(T_2^2 + T_1^2)(T_2 + T_1) \qquad (6.37)$$

6.3.2.2 Plate thermocouple (PT) control in standard fire resistance tests

The calibration factor of 0.45 proposed by Kay et al. (1996) to be used for the standard fire exposure has been obtained by retrospective analysis of fire tests on unprotected steel sections in a standard fire testing furnace where the furnace temperatures were controlled by conventional thermocouples. Because this type of thermocouples have a small thermal resistance, they are more sensitive to the convective heat transfer of the fire and may not accurately measure the radiant heat flux of the fire.

Recently, the plate thermometer (Wickstrom 1988, 1989) has been proposed and adopted in Europe to control the temperature regime in standard fire testing furnaces. The plate thermometer has a much larger area (100×100 mm) and has a near blackbody surface. Therefore, the temperature recorded by the plate thermometer may be regarded as the "effective" blackbody temperature of the fire. In this case, the emissivity of the "fire" as recorded by the plate thermometer is unity and no modification factor should be used to distinguish between different fire exposure conditions (Wickstrom and Hermodsson 1997). Equation (6.37) can still be used to calculate the radiant heat transfer coefficient, but the resultant emissivity should be replaced by the emissivity of the surface of the construction element.

6.4 SOME SIMPLIFIED SOLUTIONS OF HEAT TRANSFER

To obtain temperature distributions in a construction element exposed to fire attack, numerical procedures are generally necessary. However, for the two common cases of unprotected and protected steelwork exposed to fire attack, simple analytical solutions have been derived to enable their temperatures to be calculated quickly.

These simple analytical solutions have been derived by using the "lumped mass method", i.e. the entire steel mass is given the same temperature. The validity of this assumption depends on the rate of heat transfer within the material, i.e. its thermal conductivity and its thickness. Considering a plate totally immersed in air that experiences a sudden rise in temperature, detailed theoretical consideration (Carslaw and Jaeger 1959) suggests that provided the Biot number is less than 0.1, the plate may be assumed to have a uniform temperature distribution and the lumped mass method may be applied.

The Biot number (Bi) is defined as:

$$\text{Bi} = \frac{ht}{2k} \tag{6.38}$$

where h is the total heat transfer coefficient at the plate surface, t the plate thickness and k the plate thermal conductivity.

As an approximation, the thermal conductivity of steel may be taken as 50 W/(m.K). At a fire temperature of $T = 1000$ K and a steel temperature of 800 K and assuming a resultant emissivity of 0.5, the radiant heat transfer coefficient from equation 6.37 is 84 W/(m^2.K). Assume a convective heat transfer coefficient of 25 W/(m^2.K), giving the total heat transfer coefficient h of 109 W/(m^2.K). For Bi < 0.1, the maximum steel plate thickness can be calculated as:

$$t = 0.1 \times 2 \times 50/109 = 0.092\,\text{m} = 92\,\text{mm}.$$

In realistic applications, the steel plate thickness is usually much lower than this value. Hence, the "lumped mass method" can be used.

6.4.1 Temperatures of unprotected steelwork in fire

Figure 6.11 shows the cross-section of a steel element subject to fire attack on all sides. Assuming that the steel temperature is T_s and the fire temperature is T_{fi}, the heat balance equation may be written as:

$$V \rho C_s \frac{dT_s}{dt} = h(T_{fi} - T_s)A_s \tag{6.39}$$

where V and A_s are the volume and the exposed surface area of the steel element; ρ is the density and C_s the specific heat of steel.

The left hand side of equation (6.39) is the heat required to increase the steel temperature during an infinitesimal period of time and the right hand side is the heat input from fire to the steel element. Using a step by step approach and assuming that the time increment is small ($\Delta t < 5\,\mathrm{s}$), the steel temperature increase (ΔT_s) may be calculated using:

$$\Delta T_s = \frac{h}{\rho C_s} \frac{A_s}{V}(T_{fi} - T_s)\Delta t \tag{6.40}$$

The ratio A_s/V in equation (6.40) is often referred to as the section factor of the steel element. Equation (6.40) has been adopted in Eurocodes 3 and 4 Part 1.2 (CEN 2000b, 2001).

If the heat flux is given, the temperature in an unprotected steel element may be calculated using (Myllymaki and Kokkala 2000a,b):

$$C_s \rho \frac{1}{A_s/V} \frac{\Delta T_s}{\Delta t} = (1 - (1 - \varepsilon_s)\gamma)\dot{Q}_{flux} - h_c(T_s - T_a) - \varepsilon_{fi}\varepsilon_s\sigma(T_s^4 - T_a^4) \tag{6.41}$$

Figure 6.11 An unprotected steel section in fire.

where ε_s is the emissivity of steel, γ the fraction of the heat flux by radiation, h_c the convective heat transfer coefficient and T_a the ambient temperature. According Myllymaki and Kokkala (2000a,b), γ is about 0.8.

6.4.2   Temperatures of protected steelwork in fire

Figure 6.12, shows the cross-section of a protected steel element under fire attack. Assume that the temperatures of the steel section and the fire are T_s and T_{fi}. According to equation (6.9), the heat transfer from fire through the fire protection to the steel section is:

$$\dot{Q}_{con} = \frac{1}{1/h + d_p/k_p}(T_{fi} - T_s)A_s\Delta t \tag{6.42}$$

where t_p and k_p are the thickness and thermal conductivity of the fire protection material, h is the overall heat transfer coefficient of fire.

As a first approximation, assume that the fire protection temperature is the average of the fire temperature and the steel temperature, i.e. $T_p = 1/2(T_{fi} + T_s)$. The total heat required to increase the temperatures of the steel and the fire protection is:

$$\dot{Q}_{req} = C_s\rho_s V\Delta T_s + C_p\rho_p t_p A_p \frac{1}{2}(\Delta T_{fi} + \Delta T_s) \tag{6.43}$$

where A_p is the exposed surface area of the fire protection.

Assume that the fire protection is thin, A_p may be taken as the surface area of the steel element. Equating equation (6.43) to (6.42) gives the increase in steel temperature as:

$$\Delta T_s = \frac{(T_{fi} - T_s)A_p/V}{(1/h + t_p/k_p)C_s\rho_s\left(1 + \frac{1}{2}\phi\right)}\Delta t - \frac{1}{\left(\frac{2}{\phi} + 1\right)}\Delta T_{fi} \tag{6.44}$$

where

$$\phi = \frac{C_p\rho_p}{C_s\rho_s}t_p\frac{A_p}{V}$$

The term $1/h$ in equation (6.44) is the thermal resistance of fire. Compared to that of the insulation material, this term is small and may be ignored.

When deriving equation (6.44), a number of assumptions have been made. More detailed theoretical considerations by Wickstrom (1982, 1985) suggest that equation (6.45) may be used to give more accurate predictions of the temperature in a protected steel element:

Figure 6.12 A protected steel section in fire.

$$\Delta T_s = \frac{(T_{fi} - T_s)A_p/V}{(t_p/k_p)C_s\rho_s\left(1+\frac{1}{3}\phi\right)}\Delta t - \left(e^{\phi/10}-1\right)\Delta T_{fi} \qquad (6.45)$$

Equation (6.45) is adopted in Eurocodes 3 and 4 Part 1.2 (CEN 2000b, 2001).

The time increment should not be too large. When using equation (6.45), the time increment (Δt) should not exceed 30 s.

Because of the second term in equation (6.45), it is possible that at the early stage of increasing fire temperature, the steel temperature may decrease. In this case, the steel temperature increase should be taken as zero.

6.4.2.1 Section factors

Equations (6.40) and (6.45) clearly indicate that the temperature rise in a steel element is directly related to the section factor, i.e. the ratio of the heated surface area to the volume of the steel element. Consider a unit length of a steel element where the end effects are ignored, the section factor may alternatively be expressed as H_p/A. Here H_p is the fire exposed perimeter length of the steel cross-section and A the cross-sectional area of the steel element. Section factors for a few common types of steel sections exposed to fire are given in Table 6.5.

In Table 6.5, it is assumed that the entire cross-section of the steelwork is at the same temperature. Sometimes, e.g. in the case of unprotected steelwork, it can be more useful to obtain different temperatures in different parts of the steel cross-section. In this case, the so-called "element factor" (Kirby and Tomlinson 2000) may be used. Element factors for different parts of an unprotected universal beam/column section are given in Figure 6.13.

Table 6.5 Section factors of a steel element

Fire exposure situation	A_p/V
Unprotected steel section exposed to fire exposure around all sides	$\dfrac{2(2B - t_w + D)}{A_s}$

Fire exposure on all sides of board protection	$\dfrac{2(B + D)}{A_s}$

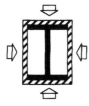

Fire protection on three sides: profile protection	$\dfrac{2(B - t_w) + B + 2D}{A_s}$

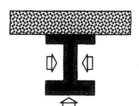

Fire protection on three sides: board protection	$\dfrac{2D + B}{A_s}$

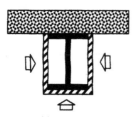

Notes
B = section width; D = steel depth; A_s = cross-sectional area; t_w = web thickness.

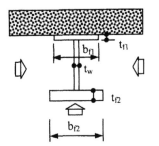

Element factors:
Upper flange: $(2t_{f1}+b_{f1}-t_w)/(t_{f1}*b_{f1})$
Web: $2/t_w$
Lower flange: $(2b_{f2}+2t_{f2}-t_w)/(b_{f2}*t_{f2})$

Figure 6.13 Element factors for an unprotected steel section.

6.4.2.2   *Temperatures in protected steelwork under natural fires*

Although equation (6.45) is adopted in Eurocodes for general purposes, it should only be applied to the standard fire exposure. Under the standard fire exposure, the fire temperature is always increasing, therefore the second term on the right hand side of equation (6.45) is always negative, representing the heat sink effect of the fire protection material. Natural fires reduce in temperature during the cooling phase and the second term on the right hand side of equation (6.45) contributes to an increase in the steel temperature. Although an increase in the steel temperature is possible during the cooling stage of a fire, it is expected that the rate of increase in the steel temperature during the cooling stage will be lower than that during the heating stage. However, it is not too difficult to find situations in which equation (6.45) will give faster increases in the steel temperature during the cooling stage than during the heating stage.

For example, Figure 6.14 shows the steel temperature response to a natural fire whose temperature–time relationship is also shown. Clearly, equation (6.45) predicts an unrealistically high rate of increase in the steel temperature during the cooling stage of fire.

To rectify this situation, Wang (2001a) carried out a detailed study and found that the rate of increase in the steel temperature during the cooling stage of a fire should be limited to half of that at the end of the heating phase. The modified temperature calculation procedure is illustrated in Figure (6.15). Using this modification, the steel temperature response can be predicted very accurately as indicated in Figure (6.14).

6.5   THERMAL PROPERTIES OF MATERIALS

In order to use equations (6.40) and (6.45), it is necessary to have available information on the thermal properties (thermal conductivity k, density ρ and specific heat C) of steel, concrete and insulation materials.

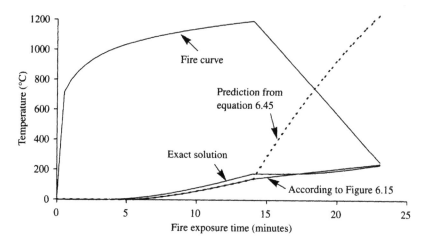

Figure 6.14 Different predictions of protected steel temperature, $\phi = 7.76$ (from Wang 2001a).

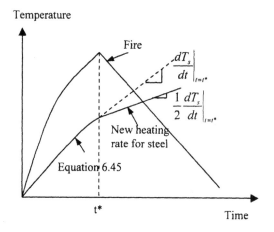

Figure 6.15 Modified method for calculating protected steel temperature.

6.5.1   Steel

The thermal properties of steel are known with reasonable accuracy and the following values are given in Eurocode 3 Part 1.2 (CEN 2000b).

Density:

$$\rho_s = 7850 \, \text{kg/m}^3$$

Thermal conductivity [W/(m.K)]:

$$k_s = 54 - \frac{T_s}{300} \quad \text{for} \quad 20\,^\circ\text{C} \le T_s \le 800\,^\circ\text{C}$$

$$k_s = 27.3 \qquad \text{for} \quad T_s > 800\,^\circ\text{C}$$

Specific heat [J/(kg.K)]:

$$C_s = 425 + 0.773T_s - 0.00169T_s^2$$
$$+ 2.22 \times 10^{-6}T_s^3 \quad \text{for} \quad 20\,^\circ\text{C} \le T_s \le 600\,^\circ\text{C}$$

$$C_s = 666 - \frac{13002}{T_s - 738} \quad \text{for} \quad 600\,^\circ\text{C} < T_s \le 735\,^\circ\text{C}$$

$$C_s = 545 - \frac{17820}{T_s - 731} \quad \text{for} \quad 735\,^\circ\text{C} < T_s \le 900\,^\circ\text{C}$$

$$C_s = 650 \qquad \text{for} \quad T_s > 900\,^\circ\text{C}$$

These data are also shown in Figures 6.16 and 6.17.

6.5.2 Concrete

There is a great variation in the thermal properties of concrete. For general information, Figures 6.16 and 6.17 give the Eurocode 4 (CEN 2001) values of thermal conductivity and specific heat for dry normal weight and lightweight concrete.

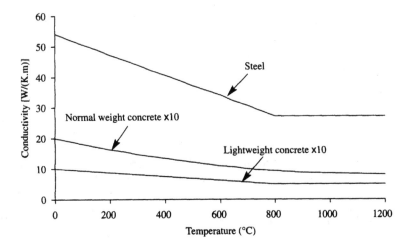

Figure 6.16 Thermal conductivity of steel and concrete (from CEN 2001). Reproduced with the permission of the British Standards Institution under licence number 2001SK/0298.

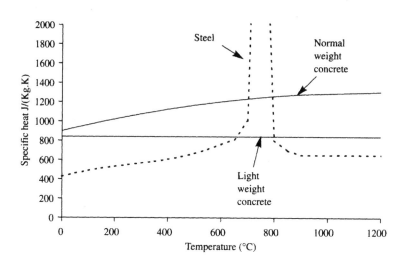

Figure 6.17 Specific heat of steel and concrete (from CEN 2001). Reproduced with the permission of the British Standards Institution under licence number 2001SK/0298.

Concrete is a hygroscopic material in that it contains water. Due to the migration of water, the heat transfer process is affected. To accurately evaluate temperatures in this type of material, a combined temperature and mass transfer analysis should be carried out. However, this will make the modelling process rather laborious. Whilst combined heat and mass transfer modelling is necessary for concrete construction to obtain moisture-induced pore pressure distributions, unless concrete spalling occurs, this is not necessary in pure temperature calculations. An approximate method is to add the energy consumed by water evaporation at about 100 °C to the specific heat of the material.

The total energy required to evaporate water in a unit weight of material is $w_c \times Q_w$, where w_c is the moisture content by weight and Q_w the latent heat of vapour $Q_w = 2260\,\text{kJ/kg}$. Assume that water does not move inside the material and that evaporation takes place within a temperature range. As shown in Figure 6.18, if increase in the specific heat of the hygroscopic material has a triangular distribution with temperature, the maximum increase in the specific heat over that of the dry material may be obtained from:

$$\Delta C = \frac{2w_c \times Q_w}{\Delta T} \tag{6.46}$$

where ΔT is the temperature range during which water evaporation occurs. As an approximation, ΔT may be taken as 50 °C, assuming water evaporation starts at 85 °C and ends at 135 °C.

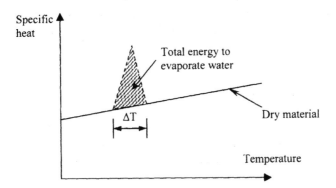

Figure 6.18 Increase in specific heat due to moisture.

6.5.3   Insulation materials

It is much more difficult to obtain information on the thermal properties of fire protection materials. This is partly due to the specific nature of fire protection materials, which have complicated and variable chemical reactions at high temperatures. An important factor contributing to this lack of information is that most fire protection materials are proprietary systems from different manufacturers and commercial sensitivity prevents publication of this type of information.

Information in Table 6.6 should only be used as a general guide for a few generic types of fire protection material.

Intumescent coatings offer a number of advantages, e.g. architectural appearance and the possibility of offsite applications, and are increasingly being used as fire protection materials to steel structures. At present,

Table 6.6 Thermal properties of generic fire protection materials

Generic material	Density kg/m³	Thermal conductivity W/(m.K)	Specific heat J/(kg.K)	Moisture content (% by wt.)
Sprayed mineral fibre	250–350	0.1	1050	1.0
Vermiculite slabs	300	0.15	1200	7.0
Vermiculite/gypsum slabs	800	0.15	1200	15.0
Gypsum plaster	800	0.2	1700	20.0
Mineral fibre sheets	500	0.25	1500	2.0
Aerated concrete	1600	0.3	1200	2.5
Lightweight concrete	1600	0.8	1200	2.5
Normal weight concrete	2200	1.7	1200	1.5

Source: Lawson and Newman (1996). Reproduced from *SCI Publication* 159, "Structural fire design to EC3 & EC4, and comparison with BS 5950", with the permission of The Steel Construction Institute. Ascot. Berkshire.

there are no suitable simplified design equations for steel structures protected with intumescent coatings. The designer has to rely on information provided by the intumescent coating manufacturer from standard fire resistance tests.

6.6   NUMERICAL ANALYSIS OF HEAT TRANSFER

Inevitably, it is difficult to obtain analytical solutions to general problems of heat transfer and numerical methods are necessary. To help the reader understand heat transfer software packages, this section will give some basic equations of 3-D transient state heat transfer.

6.6.1   Three-dimensional steady-state heat conduction

Figure 6.19 shows the heat conducting element $dx \times dy \times dz$ with its edges parallel to the Cartesian coordinates x, y, and z. To establish the basic steady state heat conduction equation in three dimensions, it is necessary to use the energy conservation principle, i.e. for steady state heat transfer, the heat inflow into the element should be equal to the heat outflow from the element.

Consider heat transfer in the x-direction. The total heat inflow to the element is:

$$\dot{Q}_{x,i} = -k_x \frac{\partial T}{\partial x} dy dz \tag{6.47}$$

The total heat outflow from the element is:

$$\dot{Q}_{x+dx,o} = -k_{x+dx} \frac{\partial \left(T + \frac{\partial T}{\partial x} dx \right)}{\partial x} dy dz = -k_{x+dx} \left(\frac{\partial T}{\partial x} + \frac{\partial^2 T}{\partial x^2} dx \right) dy dz \tag{6.48}$$

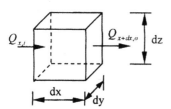

Figure 6.19 Heat conduction in a 3-D control volume.

Assuming that the thermal conductivity of the element is constant within an infinitesimal space so that $k_x = k_{x+dx}$, the net heat inflow into the control volume in the x-direction is:

$$\Delta \dot{Q}_x = \dot{Q}_{x,i} - \dot{Q}_{x+dx,o} = k_x \frac{\partial^2 T}{\partial x^2} dxdydz = k_x \frac{\partial^2 T}{\partial x^2} dV \qquad (6.49)$$

Similarly, net heat inflows into the control volume in y and z directions are:

$$\Delta \dot{Q}_y = k_y \frac{\partial^2 T}{\partial y^2} dV \qquad (6.50)$$

and

$$\Delta \dot{Q}_z = k_z \frac{\partial^2 T}{\partial z^2} dV \qquad (6.51)$$

According to the principle of energy conservation, the steady state heat conduction equation is thus:

$$k_x \frac{\partial^2 T}{\partial x^2} + k_y \frac{\partial^2 T}{\partial y^2} + k_z \frac{\partial^2 T}{\partial z^2} = 0 \qquad (6.52)$$

6.6.2 Three-dimensional transient-state heat conduction

Under transient-state heat conduction, temperatures will change with time. Referring to the control volume in Figure 6.19, the total net outflow of energy is that necessary to increase the temperature of the control volume. It is given by $[\rho C(\partial T/\partial t)] \times dV$, where ρ is the density and C the specific heat of the control volume. Equating the net heat inflow to the net heat outflow, the transient state heat conduction equation is:

$$k_x \frac{\partial^2 T}{\partial x^2} + k_y \frac{\partial^2 T}{\partial y^2} + k_z \frac{\partial^2 T}{\partial z^2} = \rho C \frac{\partial T}{\partial t} \qquad (6.53)$$

The problem of numerical heat transfer analysis is to find numerical solutions of temperatures to equation (6.53).

6.6.3 Boundary conditions for heat transfer

To find specific solutions to equation (6.53), boundary conditions must be applied. In a general heat transfer problem, there are three types of boundary condition.

For the first type, temperatures at the boundary surface are specified, i.e.

$$T = T_0|_{\text{at the boundary}} \qquad (6.54)$$

For the second type of boundary condition, the heat flux is specified. Therefore:

$$-k_n \frac{\partial T}{\partial n}\bigg|_{\text{at the boundary}} = \dot{q}_0 \qquad (6.55)$$

where n is measured in the direction normal to the boundary surface whose heat flux is $\dot{q}_0$.

For a construction element under fire conditions, the boundary condition is generally described by the third type. Here the construction element is surrounded by a fluid (fire or ambient air) and the boundary condition is expressed as:

$$-k_n \frac{\partial T}{\partial n}\bigg|_{\text{at the boundary}} = h_f \Delta T \qquad (6.56)$$

where h_f is the total heat transfer coefficient and ΔT the temperature difference between the fluid and the boundary surface of the construction element. The total heat transfer coefficient h_f consists of the two parts from convective and radiant heat transfers. Therefore:

$$h_f = h_c + h_r \qquad (6.57)$$

where the convective part h_c may be obtained from equation (6.10) and the radiant part h_f from equation (6.37).

There are many standard textbooks on numerical analysis of heat transfer (e.g. Bathe 1996) and a number of well-known computer programs for heat transfer analysis exist in the fire engineering field such as TASEF (Wickstrom 1979), TEMPCALC (IFSD 1986), FIRES-T3 (Iding et al. 1977a) and SAFIR (Franssen et al. 2000). The main problem is how to deal with the radiant boundary condition when calculating the radiant heat transfer coefficient h_r. It is usually assumed in these computer programs that fire is in contact with construction elements so that the configuration factor may be taken as 1.0 and the resultant emissivity in equation (6.35) may be used.

Numerical heat transfer analysis can also be carried out using many general commercial finite element analysis packages such as ANSYS, ABAQUS, ADINA or DIANA. Suffice to say that all these numerical methods employ similar assumptions and techniques to solve the same general problem given in equation (6.53). They also have similar accuracy. Therefore, when deciding which program to use for structural applications, availability and user-friendliness of the program should be the main decision factors.

Chapter 7

An introduction to enclosure fire behaviour

Fires can occur anywhere and have different severity. This chapter will only describe fires which take place in an enclosure. Fire development in an enclosure undergoes a number of stages. In the early stages, combustion is restricted to local areas near the ignition source and temperatures of the combustion gases are low, but there is a large quantity of smoke and toxic chemicals. It is the reduced visibility due to smoke and the toxicity of chemical products that pose the biggest threat to life. Therefore, the main focus of the fire research community has concentrated on understanding ignition, production of smoke and toxic gases, their movement and their effects on human behaviour. Since fire temperatures at this stage are usually low, the safety of a structure exposed to fire attack is very rarely threatened. Hence this aspect of fire behaviour has not attracted the attention of many structural fire engineers. Their interest is usually focused on the later stage of fire development when the fire involves all combustible materials and the combustion gases attain very high temperatures. The stability of the structure may be threatened, the consequence of which can lead to rapid fire spread and loss of life and property. This is the so-called post-flashover fire and it is this stage of the fire development that will be addressed in this chapter.

This chapter will also provide some information on localized fires that occur in large compartments (such as car parks) where it is not possible for a fire to fully involve all combustible materials, yet structural members in the vicinity of the fire can still be severely affected.

7.1 A GENERAL DESCRIPTION OF ENCLOSURE FIRE BEHAVIOUR AND ITS MODELLING

The development of an enclosure fire may broadly be divided into three stages: (1) fire growth; (2) steady burning; and (3) decay. Fire growth starts from ignition of the first burning item inside the enclosure. During this stage, the fire is localized and temperature distribution inside the enclosure is highly non-uniform. The main danger is the risk to life safety due to the production of large quantities of smoke and toxic gases. If this fire is promptly discovered and effective fire fighting is activated, it can be easily controlled and fire damage and risk will be minimal. If there is no intervention, but the first burning item is sufficiently far away from other combustible materials, the fire may die out due to the difficulty of igniting other combustible materials. Also, if there is an insufficient supply of oxygen the fire may appear to die down, but it may grow again if fresh air is supplied into the enclosure. In more dramatic situations, a sudden fresh air supply to an under-ventilated fire may lead to the so-called "backdraught" phenomenon, posing serious hazards for fire fighting.

In structural fire engineering, it is assumed that fire spread will occur and that there is a sufficient supply of fresh air to aid fire growth. During the fire growth stage, hot gases will be released from burning combustible materials. Due to buoyancy, these hot gases (fire plume) will rise towards the ceiling (see Figure 7.1a). In the meantime, fresh air is entrained and more hot smoke is released. On hitting the ceiling, the smoke will spread outwards until it hits the walls. Afterwards, a smoke layer will quickly form

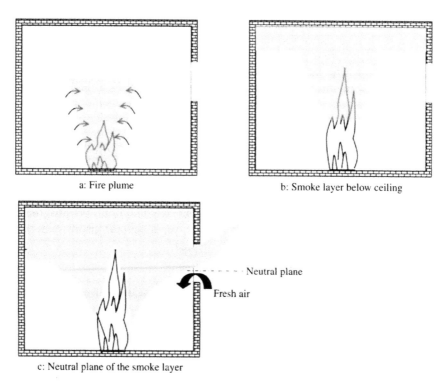

a: Fire plume

b: Smoke layer below ceiling

c: Neutral plane of the smoke layer

Figure 7.1 Different stages of pre-flashover fire development.

underneath the ceiling (see Figure 7.1b). With the production of more smoke, the smoke layer will descend as it becomes thicker. On reaching the top of any opening, e.g. an open doorway or window, smoke will flow out of the enclosure. To allow fresh air to be entrained, the smoke layer will stabilize at a height within the opening (see Figure 7.1c). At this stage, the enclosure fire environment may be approximately divided into two zones: (1) an upper zone of hot smoke; and (2) a lower zone of cold air. The division between the upper and lower zones is the neutral plane, above which smoke flows out of the enclosure and below which fresh air is supplied into the enclosure.

As the volume of smoke is stabilized while more hot combustion gases are continuously supplied into the smoke layer, the smoke temperature increases. Owing to a lack of oxygen in the smoke layer, a large quantity of partially burnt fuel will also accumulate in the smoke layer. Meanwhile, the burning flame will become larger and penetrate the smoke layer (see Figure 7.2). Flame spread becomes quicker when it is aided by the partially burnt fuels in the smoke layer. The radiation from the burning flames and the high temperature smoke layer will increase the burning rate of the existing fire. All this will accelerate a positive burning loop.

This rapidly accelerating positive burning loop will lead to an instant when the incident radiation on the unburned combustible materials in the enclosure becomes so high that they become ignited at almost the same time. The onset of such an event is often referred to as "flashover" when all combustible materials inside the enclosure become involved in fire. The fire growth stage before the onset of flashover is referred to as pre-flashover fire and that afterwards post-flashover fire. The transition from a pre-flashover to post-flashover fire occurs in a short time interval. After flashover, the fire temperature increases rapidly.

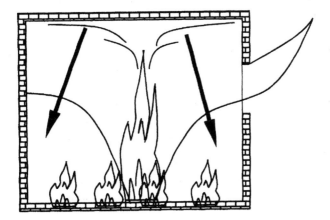

Figure 7.2 Simultaneous ignition and flashover.

After flashover, the fire enters a steady burning stage. Since all combustible materials are now burning, the rate of burning depends on the amount of fresh air that can be supplied. In other words, the fire is ventilation controlled. There are some occasions when the area covered by combustible materials in the fire enclosure is small. In these cases, the rate of burning depends on the surface area of burning fuels. This type of burning is referred to as fuel bed controlled.

During post-flashover, the burning rate of the fire reaches the maximum; the fire temperature is high and damage to the building structure reaches the maximum. This fire stage is the main concern of structural fire engineers and will be discussed in detail in the next section. After a period of sustained burning, most combustible materials will be consumed and the burning rate of the fire starts to decrease. The fire now enters the decay stage. The fire will eventually die out when all combustible materials in the fire enclosure have been consumed. Figure 7.3 illustrates the complete temperature–time relationship of an enclosure fire.

Modelling the behaviour of a fire mathematically is a very sophisticated process and is beyond the scope of this book. Suffice to say that this has attracted the attention of fire scientists all over the world. As a result of these efforts, many tools and models have been developed. For detailed information, interested readers should refer to excellent publications by Cox (1995), Drysdale (1999), Karlsson and Quintiere (1999) and Quintiere (1989a). Broadly speaking, these modelling techniques may be divided into three types (Thomas 1992):

1 analytical methods;
2 zone modelling;
3 computational fluid dynamics (CFD) modelling.

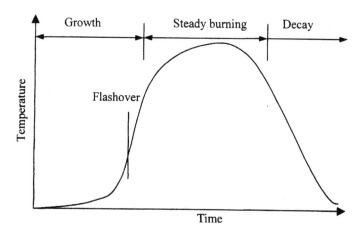

Figure 7.3 Temperature–time relationship of an enclosure fire.

Analytical methods give the simplest and most elegant analytical solutions. Here, a large number of fire tests are carried out in which different design parameters are varied. Regression analyses are then performed to fit relationships between output variables (e.g. burning rate, fire temperature, flame size) and design input variables such as fuel load, ventilation size etc. However, deducing the appropriate combinations of input and output variables and their formats for regression analysis demands a deep understanding of the physical problem. This is often done with the help of scaling analysis (Quintiere 1989b) so that limited test results can be applied more broadly. Due to complexity of the fire problem, analytical solutions can be found in only a few limited cases.

On the other hand, in CFD modelling, the fire enclosure is divided into a large number of control volumes. Partial differential equations of mass, momentum and energy transfer and conservation of species are written for each control volume based on fundamental equations of fluid dynamics, thermodynamics, chemical reactions and mechanics. They are then numerically assembled for the entire fire enclosure. Appropriate initial boundary conditions are then applied to find numerical solutions to these equations. Clearly, CFD modelling is versatile and has the potential to give the most accurate and detailed results for a wide range of problems. Inevitably, CFD modelling requires specialist knowledge on fire modelling and has so far been mainly used for predicting the production and movement of smoke and toxic gases in pre-flashover fires at a given rate of heat release of the fire. However, CFD modelling is now being extended to simulate flame spread and ignition (Yan and Holmstedt 1997).

If the above two approaches represent the two extremes of fire modelling, the middle approach is zone modelling (Quintiere 1989a). Here, a fire enclosure is divided into a few large spaces with distinctly different characteristics. For example, for pre-flashover fire modelling, the fire enclosure may be divided into two zones: (1) an upper zone of hot smoke; and (2) a lower zone of cold air. Each zone is assumed to have uniform properties, e.g. temperature and smoke concentration. Zone models are much less complicated to use than CFD models, but they give less detailed information. Compared to CFD models, they are less versatile and require pre-knowledge about the fire behaviour, e.g. how to divide the enclosure into appropriate zones and interactions between them. However, they are suited to most engineering applications where the geometry of the enclosure is not complicated and there is already a good qualitative understanding of the fire behaviour.

Zone models may be used for predicting the behaviour of both pre-flashover and post-flashover fires. For post-flashover fire modelling which is of particular interest to structural fire engineering, the entire fire enclosure may be considered as one zone at the same temperature. In Section 7.3, a one-zone model will be developed to estimate temperatures of post-flashover fires.

In fire modelling using the above-mentioned techniques, a deterministic approach is followed. Here, it is assumed that each fire can be uniquely described by a set of well-defined conditions. However, it must be recognized that they are uncertainties about many factors that affect fire behaviour, e.g. the location of fire ignition; the behaviour of glass; the wind condition outside the fire enclosure; the random nature of the opening of doors; the variability in properties of construction materials. To include all these variables into consideration would necessitate a probabilistic approach. A probabilistic approach has the potential to be more realistic and there are various attempts to use this approach in fire engineering design. However, due to a lack of information on various probabilistic distributions and the complicated nature of this process, probabilistic based models are not yet practical tools. In most engineering applications, the deterministic approach is used and the worst case is considered.

7.2 BEHAVIOUR OF LOCALIZED FIRES

Owing to the high risk with which smoke and toxic gases threaten life, the majority of studies in fire dynamics have concentrated on the pre-flashover fire so as to develop an understanding of the production and spread of smoke and toxic gases. For structural engineers, being able to predict the pre-flashover fire behaviour enables one to investigate structural behaviour under localized burning in such construction as car parks, stadia and airports. Here due to large spaces, flashover is not possible and the structure is subject to localized burning.

Although zone models may be used to simulate the behaviour of pre-flashover fires, it is possible to use some simplified analytical equations that have reasonable accuracy. Of many analytical models, those of Thomas and Hasemi have been widely adopted.

7.2.1 Thomas model of smoke temperature

As described in Section 7.1, in a pre-flashover fire, the fire enclosure may be divided into two zones: (1) an upper zone of hot smoke; and (2) a lower zone of cool air. To err on the safe side, the temperature of a structure in the upper smoke zone may be assumed to be the same as that of the hot smoke. This hot smoke temperature may be calculated according to Thomas *et al.* (1963) and Hinkley (1986).

The mass of smoke below the ceiling is approximately:

$$\dot{m}_{smoke} = 0.188 P H^{3/2} \tag{7.1}$$

where $\dot{m}_{smoke}$ is the rate of smoke production in kg/s, P is the perimeter length of the fire source (in m) and H (in m) is the ceiling height above the fire source.

Equation (7.1) is applicable to large fires and the range of application is:

$$H < 2.5P \text{ and } 200\,\text{kW/m}^2 < \dot{Q} < 750\,\text{kW/m}^2$$

where $\dot{Q}$ is the rate of heat release of the fire.

The average smoke temperature can be obtained from:

$$T_{\text{smoke}} = \frac{\dot{Q}_{\text{p}}}{\dot{m}_{\text{smoke}} C_{\text{smoke}}} + T_{\text{a}} \qquad (7.2)$$

where $\dot{Q}_{\text{p}}$ is the convective part of the total rate of heat release (kW) from the fire source, usually taken as 2/3 of the total rate of heat release. C_{smoke} is the specific heat of smoke ($\approx 1\,\text{kJ/(kg.K)}$), cf. Table 6.1.

7.2.2   Hasemi's model

Thomas' model may be used to give the average smoke temperature in the upper smoke layer other than directly above the fire source. Directly above the fire source, the ceiling temperature is much higher due to incident radiation from flames and/or direct flame impingement. To calculate the structural temperature, the model of Hasemi et al. (1997), Hasemi (1999), Myllymaki and Kokkala (2000a,b), Pchelintsev et al. (1997) for heat flux distribution may be used. In Figure 7.4, the heat flux $\dot{q}_{\text{s}}$ (in kW/m^2) to a point in the ceiling at a distance r (m) from the centre of the heat source is given by:

$$\dot{q}_{\text{s}} = 518.8 e^{-3.7x} \quad \text{with} \quad x = \frac{r + H + z'}{L_{\text{H}} + H + z'} \qquad (7.3)$$

where H is the distance between the ceiling and fire source (in m) and L_{H} is the horizontal length of the flame on the ceiling (in m) and is obtained from:

$$\frac{L_{\text{H}} + H}{H} = 2.90 Q_{\text{H}}^{*\,0.4} \qquad (7.4)$$

z' is the height of the so-called virtual burning source and is obtained from:

$$z' = 2.4D(Q^{*2/5} - Q^{*2/3}) \quad \text{for} \quad Q^* < 1.00 \qquad (7.5(a))$$

and

$$z' = 2.4D(1.00 - Q^{*2/5}) \quad \text{for} \quad Q^* \geq 1.00 \qquad (7.5(b))$$

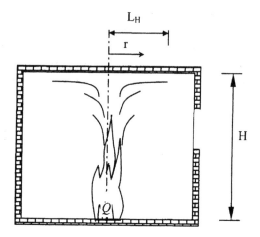

Figure 7.4 Some dimensions of a localized fire.

In equations (7.4) and (7.5), $\dot{Q}^*$ and $\dot{Q}_H^*$ are dimensionless variables related to the rate of heat release of the fire. They are defined as:

$$\dot{Q}^* = \frac{\dot{Q}}{\rho_a C_a T_a g^{1/2} D^{2.5}} = \frac{\dot{Q}}{1.11 \times 10^6 D^{2.5}} \quad \text{and} \quad \dot{Q}_H^* = \frac{\dot{Q}}{1.11 \times 10^6 H^{2.5}} \quad (7.6)$$

where $\dot{Q}$ is the rate of heat release of the fire (in watts). D is the diameter (in m) of a circular fire source. ρ_a, C_a and T_a are the density, specific heat and temperature (in K) of the cold air. For other shapes of fire source, D may approximately be taken as the diameter of a circle with equal area to the actual fire source.

Equation 7.3 may be used to calculate structural steel temperatures above a localized fire using equation (6.41) in the previous chapter. $\dot{Q}_{\text{flux}}$ in equation (6.41) should take the value of $\dot{q}_s$ in equation (7.3). When calculating structural temperatures in a localized fire, the depth of the structure

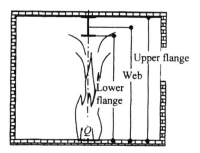

Figure 7.5 Different heights to be used for H in equations (7.3), (7.4) and (7.6).

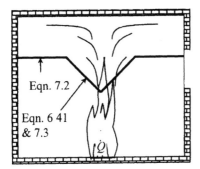

Figure 7.6 Design steel temperature in a localized fire.

should be considered. For the common case of an I beam exposed to a localized fire, Figure 7.5 illustrates the distance H that should be used in equations (7.3)–(7.6) for different parts of the cross-section.

Hasemi's equations may be applied to localized fires in an unconfined space. In a confined space, due to smoke accumulation, the smoke temperature not directly above the fire may be higher than that calculated using equations (6.41) and (7.3). In this case, the higher value from equations (6.41) and (7.2) should be used (CEC 1997a,b). Figure 7.6 illustrates this procedure.

7.3 POST-FLASHOVER FIRE MODELLING

Assuming that a fire enclosure is at the same temperature, the objective of post-flashover fire modelling is to obtain the fire temperature–time relationship so that the severity of thermal onslaught on a structure in the fire enclosure may be evaluated. The post-flashover fire temperature–time relationship may be determined by carrying out an energy balance analysis for the fire enclosure, i.e.

heat input into the fire (heat released from combustion)

= heat losses from the fire
$$(7.7)$$

In Figure 7.7, heat losses from the fire include:

1   heat lost to the outside by convection of hot gases flowing out of the fire compartment ($\dot{Q}_{lc}$);
2   heat lost to the enclosure lining ($\dot{Q}_{lw}$);
3   heat lost to the outside environment by radiation through the opening ($\dot{Q}_{lr}$);
4   heat required to increase the combustion gas temperature ($\dot{Q}_{lg}$).

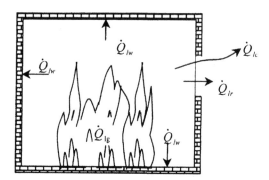

Figure 7.7 Fully developed enclosure fire, showing various heat losses.

Thus, if the heat release rate of fire is $\dot{Q}_{fi}$, the energy balance equation is:

$$\dot{Q}_{fi} = \dot{Q}_{lc} + \dot{Q}_{lw} + \dot{Q}_{lr} + \dot{Q}_{lg} \tag{7.8}$$

Each term in equation (7.8) is now discussed in turn.

7.3.1   Rate of heat release in fire ($\dot{Q}_{fi}$)

It is obvious that the rate of heat release in fire is the most important parameter to be determined in post-flashover fire modelling. Unfortunately, accurate quantification of this parameter based on understanding of the fundamental behaviour of the combustion of materials is not available and empirical equations have to be used. Over many years, a number of empirical equations have been developed for the rate of heat release in enclosure fires where wood based materials are the main combustibles. Of these, the following few are better known.

7.3.1.1   Kawagoe's equation (Kawagoe 1958)

Assuming ventilation-controlled post-flashover fire and stoichiomatic consumption of oxygen by wood based fuel, Kawagoe (1958) obtained the following equation for the burning rate of wood cribs in a post-flashover enclosure fire:

$$R = kA_v\sqrt{h} \tag{7.9}$$

where R is the burning rate of wood (in kg/min); A_v the vertical opening area (in m²); h the vertical opening height (in m) and k is an empirical constant with a value between 5 and 6 kg/(min.m$^{5/2}$), often taken as 5.5.

Whilst equation (7.9) can be shown to be a very crude approximation to experimental results, Kawagoe was the first to identify the importance of the ventilation factor $A_v\sqrt{h}$.

7.3.1.2 CIB equation

Based on the results of a large number of fire tests in small compartments (Thomas and Heselden 1972) carried out in the 1960s and 1970s coordinated by the Conseil International du Batiment (CIB), the following equation was proposed:

$$R = 1.2\left(A_T\frac{W}{D}A_v\sqrt{h}\right)^{1/2}$$

(7.10)

where A_T is the total area of the fire enclosure (in m^2), W (in m) and D (in m) are the width (containing the opening) and depth of the fire enclosure.

7.3.1.3 Law's equation

Based on the same results of CIB fire tests, Law proposed the following alternative equation (Law and O'Brien 1989):

$$R = 10.8\left\{1 - e^{-0.036\frac{A_T}{A_v\sqrt{h}}}\right\}A_v\sqrt{h}\left(\frac{W}{D}\right)^{1/2}$$

(7.11)

7.3.1.4 Harmathy's modification of Kawagoe's equation

In equations (7.9)–(7.11), there is no distinction between ventilation controlled and fuel bed controlled fires. Harmathy (1972, 1980, 1982, 1987), Harmathy and Mehaffey (1987) suggested that the burning rates of these two different types of fires would be different and would have to be considered separately. He suggested that for ventilation controlled fires, equation (7.9) of Kawagoe (1958) could still apply. For fuel bed controlled fires, the following equation should be used:

$$R = 0.372A_{ff}$$

(7.12)

where A_{ff} is the surface area of the fuel bed (in m^2).

Although there is some validity in the Harmathy approach, it is not an easy task to determine the surface area of a fuel bed in realistic applications when combustible materials are scattered.

The division between ventilation controlled and fuel bed controlled fires is checked by using the following conditions:

For fuel bed controlled fires: $5.5A_v\sqrt{h_v} \geq 0.372A_{ff}$ (7.13(a))

or $\dfrac{A_v\sqrt{h}}{A_{ff}} \geq 0.0678\,\mathrm{m}^{1/2}$;

For ventilation controlled fires: $\dfrac{A_v\sqrt{h}}{A_{ff}} < 0.0678\,\mathrm{m}^{1/2}$ (7.13(b))

7.3.1.5 Equation of Thomas and Smith

Equations (7.9)–(7.12) have been obtained based on the results of fire tests in small fire enclosures. Fires in large enclosures behave differently due to a variety of reasons such as non-uniform burning and incomplete combustion. Based on the results of fire tests by Hagen and Haksever (1986), Hagen (1991), Thomas and Smith (1993) suggested the following equation:

$$R = 2.5A_{ff}\left(\frac{A_v\sqrt{h}}{A_T}\right)^{1/4}$$ (7.14)

7.3.1.6 A comparison of different methods

Figure 7.8 compares predictions using equations (7.9)–(7.14) against some measured rates of burning (Connolly *et al.* 1995). The experimental sources

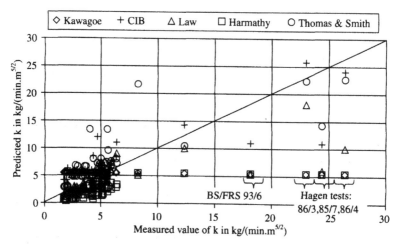

Figure 7.8 Comparison between test results and different predictions for burning rate of fire (from Connolly *et al.* 1995).

include the fire tests by Corus and the Fire Research Station in medium size compartments (Latham *et al.* 1986) and in large size compartments (Kirby *et al.* 1993), and some fire tests by Hagen (1991). Obviously, the experimental rate of burning cannot be determined very precisely. For this comparison, it was taken as the average value during sustained burning, defined as the period of weight loss from 30–80% of the total fire load. For convenience of comparison, k in Figure 7.8 is defined as:

$$k = \frac{R}{A_v\sqrt{h}}$$

(7.15)

It is clear that k is not a constant as suggested by equation (7.9) of Kawagoe. Equations (7.12) and (7.13) of Harmathy consistently underestimate the rate of burning. On the other hand, predictions using equation (7.14) of Thomas and Smith are far too high in most cases. Equation (7.10) of CIB and equation (7.11) of Law give consistently better results. Figure 7.9 is extracted from Figure 7.8 to show correlation between the test results and predictions by equations (7.10) and (7.11) of CIB and Law. It appears that the CIB method gives slightly better results for very high rates of burning ($k > 15$). However, Law's method gives much better correlations for low to medium values of k (<15). Of the four values of $k > 15$, one was well predicted by Law's equation. For the other three high values of k, it was found that the fire enclosure had extremely small openings in comparison to the enclosure surface area ($A_v\sqrt{h}/A_T$). This ratio was about $0.003\,\text{m}^{1/2}$ for the BS/FRS test 93/6 and the Hagen test 86/4 and about $0.006\,\text{m}^{1/2}$ for the Hagen test 85/7. It is very unlikely that such small openings will be encountered in practice. Overall, Law's method (equation (7.11)) may be recommended.

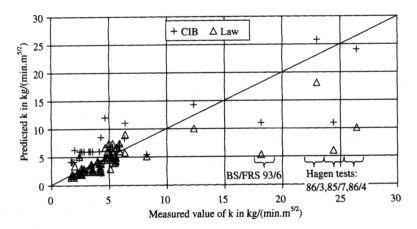

Figure 7.9 Comparison between test results and predictions of Law and CIB.

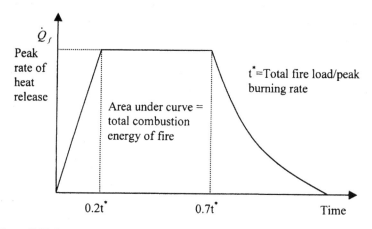

Figure 7.10 Rate of heat release–time relationship.

Once the equivalent burning rate of wood is obtained, the rate of heat release of a wood based fire is given by

$$\dot{Q}_{fi} = \alpha R \times \Delta H_c \tag{7.16}$$

where ΔH_c is the heat of combustion of wood (≈ 18.8 MJ/kg) and α is a factor used to account for burning efficiency, typically $\alpha = 0.7$.

It should be noted that equations (7.9)–(7.14) only give the maximum burning rate of wood during the steady burning period of a fire. To enable prediction of the fire temperature–time relationship over the entire period, the rate of heat release–time relationship should be obtained. This relationship is difficult to quantify. If there is no better information available on the rate of heat release–time relationship, the relationship in Figure 7.10 may be used (Wang *et al.* 1996). Here, it is assumed that 10% of the total combustible material is consumed before the maximum rate of heat release is achieved. During the steady burning stage, about 50% of the fuel is consumed. Decay follows the end of steady burning with a parabolic reduction in the rate of heat release.

7.3.2 Heat loss due to hot gas leaving fire enclosure $(\dot{Q}_{lc})$

In a fully developed fire, it is noticeable that hot gases flow out of the fire enclosure from the upper part of the opening and cold air flows into the fire enclosure through the lower part, see Figure 7.11. The position where there is no air movement is the neutral plane. At the neutral plane, the pressure of

Figure 7.11 A post-flashover fire, showing hot smoke movement (from Thomas and Law 1974). Published courtesy of BRE.

the hot gas inside the fire enclosure is in equilibrium with the pressure of cold air outside the fire enclosure. The position of the neutral plane may be obtained by using Bernoulli's equation. This procedure starts with the determination of pressure distributions inside and outside the fire enclosure as functions of the opening height. The pressure difference is then converted into an air movement velocity, which then gives the rates of mass transfer across the opening. The gain in the combustion gas mass in the fire enclosure consists of the cold air inflow and the weight of pyrolyzed combustible materials. The mass loss from the fire enclosure is the hot gas outflow. Ignoring the weight loss of combustible materials (which is small in comparison with the masses of air movement), mass balance of the combustion gases in the fire enclosure requires the cold air inflow and hot gas outflow to be equal.

Once the position of the neutral plane is determined, the amount of cold air inflow and hot gas outflow can be determined. Following this procedure, the rate of cold air inflow ($\dot{m}_{air}$) can be obtained from (Karlsson and Quintiere 2000).

$$\dot{m}_{air} = \frac{2}{3} A_v \sqrt{h} C_d \rho_a \sqrt{2g} \sqrt{\frac{1 - \rho_{fi}/\rho_a}{\left[1 + (\rho_a/\rho_{fi})^{1/3}\right]^3}} \tag{7.17}$$

where g is the gravitational acceleration, ρ_a and ρ_{fi} are densities of the cold air and hot gas respectively. C_d is the orifice (discharge) coefficient of the opening, depending on friction between air movements at the opening.

C_d is about 0.6–0.7 (Karlsson and Quintiere 2000) and $g = 9.81$ m/s^2. The ratio of ρ_a/ρ_{fi} lies within 1.8 and 5.0. ρ_0 is approximately 1.2 kg/m^3. Substitution of these values into equation (7.17) gives

$$\dot{m}_{air} = 0.5 A_v \sqrt{h} \ (\text{kg/s}) \tag{7.18}$$

Thus, the heat loss by hot gas flowing out of the fire enclosure is

$$\dot{Q}_{lc} = \dot{m}_{air} C_{fi}(T_{fi} - T_a) \tag{7.19}$$

where C_{fi} is the specific heat of the combustion gas (≈ 1 J/kg.K) and T_a the cold air temperature.

It is interesting to observe that the stoichiomatic ratio of wood is 5.7, i.e. complete burning of 1 kg of wood requires 5.7 kg of fresh air. Thus, if wood is burning at its stoichiomatic ratio, the burning rate according to the air flow rate is

$$\dot{m}_{wood} = \frac{\dot{m}_{air}}{5.7} \approx 0.09 A_v \sqrt{h} \ (\text{kg/s}) \tag{7.20}$$

The burning rate of wood obtained above is very similar to that given by Kawagoe in equation (7.9). Kawagoe's equation is indeed based on this reasoning.

7.3.3   Heat loss to the enclosure wall ($\dot{Q}_{lw}$)

Assuming that the average temperature of the enclosure wall surface is T_w, the heat loss from the fire to the enclosure wall is

$$\dot{Q}_{lw} = A_T h_w(T_{fi} - T_w) \tag{7.21}$$

where h_w is the overall heat transfer coefficient between the combustion gases and the wall, including both convection and radiation. If T_w is known, h_w can be calculated using equation (6.57) in Section 6.6.3.

Equation (7.21) cannot be quantified because of the unknown wall temperature T_w. To determine the fire temperature, equation (7.21) has to be used in conjunction with heat conduction through the enclosure lining.

In Figure 7.12, assume that a compartment wall is made of n layers of equal thickness Δx with a temperature T_i at the centre of the ith layer. Also assume that both the time increment Δt and the thickness Δx are small. The heat balance equation for the ith interior layer is

$$\rho C_i \frac{\Delta T_i}{\Delta t} \Delta x = \frac{T_{i-1} - T_i}{\frac{\Delta x}{2k_{i-1}} + \frac{\Delta x}{2k_i}} - \frac{T_i - T_{i+1}}{\frac{\Delta x}{2k_i} + \frac{\Delta x}{2k_{i+1}}} \tag{7.22}$$

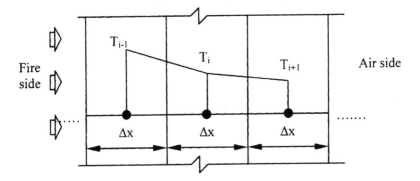

Figure 7.12 Heat conduction in a fire enclosure wall.

The left-hand side of equation (7.22) is the heat consumed to increase the temperature of the ith layer. The right hand side expresses the difference in the conducted heat into the layer and out of the layer.

The heat balance equation for the layer on the fire side is

$$\rho C_1 \frac{\Delta T_1}{\Delta t}\Delta x = \frac{T_{fi} - T_1}{\dfrac{1}{h_f} + \dfrac{\Delta x}{2k_1}} - \frac{T_1 - T_2}{\dfrac{\Delta x}{2k_1} + \dfrac{\Delta x}{2k_2}} \tag{7.23}$$

The heat balance equation for the layer on the air side is

$$\rho C_n \frac{\Delta T_n}{\Delta t}\Delta x = \frac{T_{n-1} - T_n}{\dfrac{\Delta x}{2k_{n-1}} + \dfrac{\Delta x}{2k_n}} - \frac{T_n - T_a}{\dfrac{\Delta x}{2k_n} + \dfrac{1}{h_a}} \tag{7.24}$$

In equations (7.22)–(7.24), C_i, k_i are the specific heat and thermal conductivity of the ith layer at temperature T_i; h_f and h_a are the overall heat transfer coefficients (including both radiation and convection) on the fire side and air side respectively. In total, n such equations can be written for n layers of the enclosure lining. If the fire compartment is made of different materials in the walls, the floor and the ceiling, equations (7.22)–(7.24) may be repeated for each panel.

7.3.4 Heat loss to the cold air by radiation through opening ($\dot{Q}_{lr}$)

Assume that the cold air outside is a blackbody surface, the heat loss by radiation through the opening is

$$\dot{Q}_{lr} = A_v \varepsilon_{fi} \sigma (T_{fi}^4 - T_0^4) \approx A_v \varepsilon_{fi} \sigma T_{fi}^4 \tag{7.25}$$

where ε_{fi} is the emissivity of the combustion gas and σ the Stefan–Boltzmann constant. The emissivity of the combustion gas may approximately be calculated from

$$\varepsilon_{fi} = 1 - e^{-KL_{fi}} \qquad (7.26)$$

where K is the emission coefficient and L_{fi} the gas beam length (which may be taken as the depth of the fire enclosure). According to Pettersson *et al.* (1976), $K = 1.1\,\mathrm{m}^{-1}$.

7.3.5 Heat required to increase the fire temperature ($\dot{Q}_{lg}$)

This term is simply given as

$$\dot{Q}_{lg} = \rho_f V C_a \times \Delta T_{fi} \qquad (7.27)$$

where V is the enclosure volume and ΔT_{fi} the increase in combustion gas temperature. The contribution of equation (7.27) to the total heat loss in fire is very small and may be neglected.

To summarize, provided the rate of heat release of the fire is known, there are $n + 1$ variables (n wall temperatures and the fire temperature) and $n + 1$ equations (7.8) and (7.22)–(7.24). These equations can be solved, giving the required fire temperature T_{fi}.

7.3.6 Some approximate temperature–time relationships for post-flashover fires

If accurate information on the rate of heat release–time relationship of an enclosure fire is available, the procedure described in Section 7.3 may be used to provide quite accurate results for the temperature–time relationship of the fire. However, the solution procedure involves tedious calculations of transient heat transfer through the fire enclosure and numerical methods are usually necessary. In addition, it is difficult to obtain accurate information on the rate of heat release–time relationship of a fire. For fire resistant design purpose, a number of approximate relationships for the fire temperature–time history have been developed.

7.3.6.1 Pettersson et al.

In the first comprehensive treatise on fire engineering of steel structures, Pettersson *et al.* (1976) presented temperature–time relationships for realistic fire exposures. They used the approach as described in Section 7.3. They

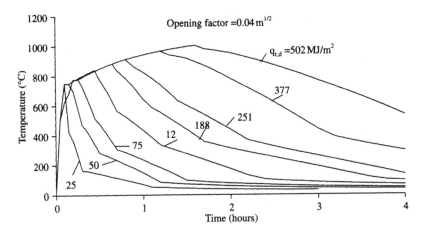

Figure 7.13 Some results of temperature–time relationships (from Pettersson et al. 1976). Reproduced with the permission of Lund University.

Figure 7.13 Some results of temperature–time relationships (from Pettersson *et al.* 1976). Reproduced with the permission of Lund University.

assumed that realistic fires were ventilation controlled so that the maximum rate of heat release of a fire was that given by equation (7.9). They then made some assumptions on the rate of heat release–time relationship of the fire. Their results are given in tables and figures, expressing the temperature–time relationship of a fire as a function of the fire load, ventilation condition and thermal properties of the fire enclosure lining.

Some results of Pettersson *et al.*, are shown in Figure 7.13 for different fire loads and ventilation factors of a "standard fire compartment". The "standard fire compartment" is constructed using lining materials with a thermal conductivity of 0.8 W/(m.K) and thermal capacitance (ρC) of 1700 kJ/(m^3.K). For other types of compartment construction, the ventilation factor and fire load are multiplied by a modification factor.

7.3.6.2 Eurocode I Part 1.2

Based on the results given by Pettersson *et al.* the European code Eurocode 1 Part 1.2 (CEN 2000a) gives a set of equations to evaluate the temperature–time relationships of post-flashover enclosure fires. These temperature–time relationships are called parametric curves. As shown in Figure 7.14, a parametric fire curve has an ascending branch and a descending branch. The ascending branch is used to describe the temperature–time relationship of a fire during its growth and steady burning stages, when it is ventilation controlled. The descending branch describes the decay period of the fire. The ascending branch is expressed by

$$T_{fi} = 1325\left(1 - 0.324e^{-0.2t*} - 0.204e^{-1.7t*} - 0.472e^{-19t*}\right) \tag{7.28}$$

This equation gives the same result as the standard heating curve (equation (7.36)), but the modified time t^* (in h) is used. The modified time is related to the real time t (in h) by

$$t^* = t\Gamma \tag{7.29}$$

where Γ is a dimensionless parameter. For the standard fire compartment of Pettersson *et al.* with a ventilation factor $O = 0.04\,\text{m}^{1/2}$, the parametric fire curve coincides with the standard fire curve and $\Gamma = 1$. For other conditions

$$\Gamma = \left(\frac{O}{0.04}\right)^2 \left(\frac{1160}{b}\right)^2 \tag{7.30}$$

In equation (7.30), O is the ventilation factor defined as

$$O = \frac{A_v\sqrt{h_v}}{A_t} \tag{7.31}$$

For a fire enclosure with different opening heights, according to Pettersson *et al.* an average height may be obtained using

$$h_v = \frac{\sum A_{vi}h_{v,i}}{\sum A_{vi}} \tag{7.32}$$

In equation (7.30), $b = \sqrt{k\rho C}$ [in $\text{J/(m}^2\text{s}^{1/2}\text{K)}$] is the overall thermal property of the fire enclosure lining material. For fire enclosures constructed of a combination of different lining materials, complicated equations have recently been introduced in Eurocode 1 Part 1.2 (CEN 2000a) to find an

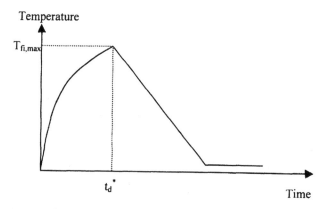

Figure 7.14 Parametric time–temperature curve of Eurocode 1 Part 1.2. Reproduced with the permission of the British Standards Institution under licence number 2001SK/0298.

equivalent value of b. Interested readers should consult the code and the background paper by Franssen (1998).

The ascending branch of the fire temperature–time relationship terminates at the time when the maximum fire temperature is obtained. This time (t_d^*) is a function of the fire load in the fire enclosure and is given by

$$t_d^* = 0.00013 \frac{q_{t,d}\Gamma}{O} \text{ (in h)} \tag{7.33}$$

In equation (7.33), $q_{t,d}$ is the fire load density (in MJ/m^2) related to the total surface area of the fire enclosure A_t. Since fire load density is usually specified with regard to the floor area A_f, the fire load per enclosure area $q_{t,d}$ is related to the fire load per floor area ($q_{f,d}$) using

$$q_{t,d} = q_{f,d}A_f/A_t \tag{7.34}$$

It can be seen that the ascending branch of the fire temperature–time curve is not dependent on the fire load. This is because a fire is assumed to be ventilation controlled and the rate of heat release is the same, depending only on the ventilation condition. The effect of fire load is to change the duration of burning t_d^* according to equation (7.33).

For simplicity, the descending branch is given by a straight line. Since structural behaviour is only slightly affected by the descending branch of the fire temperature–time relationship, it is not necessary to use complicated equations for the descending branch. The rate of the descending branch depends on the fire duration. The fire temperature during cooling is given by

$$
\begin{aligned}
T_{fi} &= T_{fi,max} - 625\left(t^* - t_d^*\right) & &\text{for } t_d^* \le 0.5; \\
T_{fi} &= T_{fi,max} - 250\left(3 - t_d^*\right)\left(t^* - t_d^*\right) & &\text{for } 0.5 < t_d^* < 2.0; \\
T_{fi} &= T_{fi,max} - 250\left(t^* - t_d^*\right) & &\text{for } t_d^* \ge 2.0.
\end{aligned}
\tag{7.35}
$$

In equation (7.35), $T_{f,max}$ is the maximum fire temperature, obtained by substituting the time in equation (7.33) into equation (7.28).

In Eurocode 1 Part 1.2 (CEN 2000a), the limit of application of the above fire temperature–time relationship is for compartments up to 100 m^2 with the maximum compartment height at 4 m. For larger or taller compartments, the effect of non-uniform temperature distribution in the fire enclosure cannot be ignored. Unfortunately, simple methods are not yet available.

In addition to the models of Pettersson et al. and Eurocode 1 Part 1.2, a number of other models have also been proposed. These models have reasonable accuracy, but their main advantage is that they are simpler to use and may be employed in quick calculations. These models include those of Babrauskas (1981), Lie (1974), Ma and Makelainen (2000) and Wang et al. (1996).

Table 7.1 Design fuel load density

Fuel load density band	Type of construction	Fuel load density $q_{f,d}$ (MJ/m^2)
I	Assembly, low hazard open sided car parks	350
II	Assembly, ordinary hazard residential office Industrial, low hazard Storage, low hazard	500
III	Shops and commercial Assembly, high hazard	750
IV	Industrial, high hazard	1000
V		1250
VI	Storage, high hazard	1500

Source: BSI (2001). Reproduced with the permission of the British Standards Institution under licence number 2001SK/0298.

7.3.7 Fire load and compartment lining properties

From previous discussions, it is clear that the fire temperature–time relationship depends on the amount of combustible materials (or fuel load) in the fire enclosure, the ventilation condition and thermal properties of the fire enclosure lining. During a design, the ventilation condition and thermal properties of the fire enclosure lining may be estimated from construction details, i.e. the window size and construction materials. Properties of some enclosure lining materials may be found in Table 6.6.

The design fire load is building specific. However, since the exact type and amount of combustible materials will not be known during the design stage, it is unlikely that the design fire load can be known with any certainty. In fire engineering design calculations, it is common to specify a generic fire load for a type of building, depending on its proposed use. This is similar to specifying a general structural load for structural design at ambient temperature. Values in Table 7.1 may be used as a guide. More detailed information on fire load may be obtained from a CIB report (CIB 1986).

7.3.8 Standard fires

Although there is a trend to move towards performance based design for fire safety of structures, current designs are still based on achieving a standard fire rating. In the standard fire resistance design, the fire temperature is calculated using the following equation (BSI 1987a)

$$T_{fi} = T_a + 345 \log(8t + 1) \tag{7.36}$$

where the standard fire exposure time (t) is in minutes.

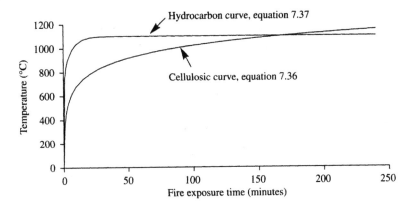

Figure 7.15 A comparison of standard cellulosic and hydrocarbon fire temperature–time relationships.

Equation 7.36 is used for wood based fires or cellulosic fires. For fire resistant design of offshore structures, the standard hydrocarbon fire curve should be used. This fire has a much faster rate of initial increase in temperature. The standard hydrocarbon fire temperature–time relationship is given by (BSI 1987a):

$$T_{fi} = 1080\left(1 - 0.325e^{-0.167t} - 0.675e^{-2.5t}\right) + T_a \qquad (7.37)$$

Figure 7.15 plots the two standard fire curves. It is obvious that the standard cellulosic fire curve gives a monotonically increasing temperature–time relationship that cannot be sustained in any real fire. In order to reflect some reality in the standard fire exposure, a limiting time of fire exposure is specified. This is the familiar standard fire rating. In standard fire resistance design calculations, specifications for the required standard fire resistance rating are based on very broad criteria such as the type and height of the building (ADB 2000; Powell-Smith and Billington 1999). Whilst these criteria give a broad indication of the fire load and consequence of fire exposure, they do not consider other important factors that affect the behaviour of an enclosure fire such as ventilation condition and construction materials.

7.3.9 Equivalent fire times

Despite having many limitations (cf. Section 3.2.2), using the standard fire exposure also has some advantages over using realistic fire exposures.

- The standard fire exposure concept has a long history and is familiar to those concerned with fire safety. The fact that building structures rarely fail in fires as a result of inadequate specification of the standard fire resistance time indicates that it is safe to use the standard fire curve.
- A large body of knowledge has been obtained from many years of standard fire resistance tests but little data exists for realistic fires.
- The standard fire curve has only one temperature–time relationship. In design calculations, it is much easier to deal with only one standard fire curve than with an infinite number of real fire curves.

Because of these advantages, attempts have been made to link realistic fires to the standard fire through the use of equivalent time. The equivalent time of a realistic fire is the time of exposing a construction element to the standard fire that would give the same effect as the realistic fire. It is important to select the appropriate "effect" to be compared. Using the temperature rise in a construction element as example, the equivalent time concept is illustrated in Figure 7.16.

Over many years, a number of relationships for the equivalent time of a realistic fire have been developed. The well-known ones include those of Harmathy (1987) and Law (1970). Eurocode 1 (CEN 2000a) also contains an equation for the equivalent time.

Law's equation was developed based on temperatures attained in insulated steelwork. It is given by

$$t_{eqv} = K' \frac{L_f}{\sqrt{A_v A_T}} \tag{7.38}$$

where L_f is the total fire load (kg wood) and K' is a constant whose value is close to unity. The equivalent time t_{eqv} is given in minutes.

In Eurocode 1 (CEN 2000a), the equivalent time is given by

$$t_{eqv} = q_{f,d} k_b w_f k_c \tag{7.39}$$

where $q_{f,d}$ is the design fire load density (MJ/m^2) per floor area, and w_f is the ventilation factor. In small compartments without horizontal opening, this is given by

$$w_f = \frac{A_f}{A_t} \frac{1}{\sqrt{O}} \tag{7.40}$$

k_b is a conversion factor to account for the influence of fire enclosure linings. O is defined in equation (7.31). Values of k_b are given in Table 7.2. k_c in equation (7.39) is a recent modification and is intended to account for different types of construction. For protected steelwork, $k_c = 1.0$. For unprotected steelwork, $k_c = 13.70$.

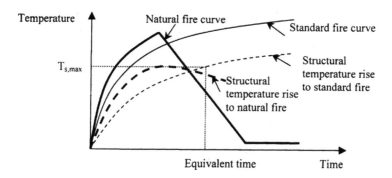

Figure 7.16 The equivalent time concept.

Harmathy's method was originally derived to calculate temperatures in reinforced concrete structures at the likely position of the reinforcement. The equivalent time in hours is given by

$$t_{eqv} = 0.11 + 0.16 \times 10^{-4} H' + 0.13 \times 10^{-9}(H')^2 \qquad (7.41)$$

Here H' is the normalized heat load of a realistic fire and is obtained from

$$H' = 10^{-6} \frac{11\delta + 1.6L_f}{A_T\sqrt{k\rho C} + 935\sqrt{\Phi L_f}} \qquad (7.42)$$

in which

$$\delta = \begin{cases} 0.79\sqrt{h_c^3/\Phi} \\ 1 \end{cases} \text{ whichever is less} \qquad (7.43)$$

Φ (kg/s) is the ventilation parameter defined as

$$\Phi = \rho_a A_v \sqrt{gh_v} \qquad (7.44)$$

and h_c (m) is the height of the fire enclosure.

Table 7.2 Values of k_b for calculating the equivalent time of a real fire

$b = \sqrt{k\rho C}$ [J/(m²s^{1/2}K)]	k_b (min.m²/MJ)
$b > 2500$	0.04
$720 \leq b \leq 2500$	0.055
$b < 720$	0.07

All equivalent time equations have been developed by comparing the time to reach a particular end temperature in a specific type of construction element. Whilst equation (7.38) of Law only considers fire load and ventilation condition, other equations also include the thermal properties of the fire enclosure lining materials.

Due to simplicity, using the equivalent standard fire rating to represent a realistic fire is attractive to fire safety engineers. However, the present author is doubtful whether the equivalent time equation obtained for one specific situation can be extended to other more general conditions. The following points are particular concerns with regards to extrapolating the equivalent time method:

- These methods have been developed by comparing temperatures at one location in a construction element. Although in some situations such as a steel structural member with uniform heating throughout, only one temperature is necessary to completely describe its behaviour, in many cases, one temperature is not sufficient for adequate representation of non-uniform temperature distributions. Examples where non-uniform temperature distributions should be considered include columns with blocked-in webs, slimfloor beams, connections and composite construction in general.

- A comparison of different equivalent time equations will show that there is a large difference in results obtained from these different methods. Although this reflects inherent differences in derivations of these methods, there is also the important question of whether the equivalent time approach that has been obtained for one type of construction element can be applied to different ones. Consider two fires: (1) having a high maximum temperature but short duration; and (2) the other with a moderate maximum temperature but long duration. It is to be expected that for a structural member that responses to fire attack quickly, such as an unprotected steel member, its maximum temperature will be close to that of the fire, thus the first fire will be more severe than the second one. On the other hand, for a structural member that has a slow response to fire attack, such as a composite member, the second fire will be able to heat the structural member to near the maximum fire temperature, but the first fire will be too short to have much effect on the structural member. Therefore, the second fire will be more severe than the first one. Hence, depending on the type of the structure, the equivalent time will be different. Although the recent addition (k_c in equation (7.39)) in Eurocode 1 Part 1.2 attempts to take this into account, it is not difficult to encounter situations where conditions in the code cannot be satisfied so that the equivalent time approach cannot be used.

In conclusion, the concept of equivalent time is an attractive proposal, being easy to use and able to reflect some of the main effects of realistic fires.

However, due to fundamental shortcomings of the approach, its limits of application should be clearly understood and the equivalent time equations should be used with extreme caution.

7.4   CONCLUDING REMARKS

The behaviour of enclosure fires is complex and better understanding is continuously being achieved. The main problem for a post-flashover enclosure fire, which is the main concern of structural fire engineers, is to obtain accurate information on the rate of heat release of the fire. At present, this information is based on empirical equations. It is difficult to progress the state of art until there is better understanding of the combustion behaviour in the fire process than at present.

Chapter 8

Design of steel structures for fire safety

Since the first comprehensive treatment of fire engineering of steel structures by Pettersson *et al.* (1976), tremendous progress has been achieved in the last 30 years or so in developing appropriate design methods for steel structures in fire. The objective of Chapters 8 and 9 is to describe how design calculations may be carried out for steel and composite steel–concrete structures. These two chapters aim to furnish the reader with an understanding of how different design methods reflect the behaviour of steel structures in fire so that the reader can develop an appreciation of the applicability and limits of these design methods. Structural behaviour in fire is a complex issue and new under-standings are continuously being obtained. Wherever appropriate, these two chapters will indicate how existing design methods may be improved or new design methods may be developed to take advantage of these new advances.

At present, the British Standard BS 5950 Part 8 (BSI 1990b, hereafter to be referred to as Part 8) and Eurocode 3 Part 1.2 (CEN 2000b, hereafter to be referred to as EC3) are the main sources of information for FR design of steel structures. Since these two design methods have many different fea-tures, both methods will be introduced. Wherever possible, design methods in these two codes will be compared and recommendations will be made as to how inconsistency between them may be reduced.

When presenting the FR design methods in Part 8 and EC3, frequent reference will be made to their parent documents BS 5950 Part 1 (BSI 1990a) and Eurocode 3 Part 1.1 (CEN 1992a) for the design of steel elements at ambient temperature. Although Chapter 2 has provided a brief introduction to these two design codes, readers should consult the original documents for more detailed information.

8.1 BASIS OF DESIGN

8.1.1 Scope of design calculations

The objectives of a FR structure are to prevent fire spread from the place of fire origin to adjacent spaces and to provide a stable structure for safe escape. This forms part of the compartmentation strategy in the fire safety design of a building. Since fire spread may occur due to failure of any of the three FR criteria (cf. Figure 3.2), it follows that all three modes of fire spread should be prevented. However, this book will only address the load bearing condition. As far as the insulation and integrity conditions are concerned, it is worth mentioning the following considerations:

- The integrity of a building compartment may be breached in two ways: (1) integrity failure of a construction element; or (2) integrity failure at junctions between different construction elements. Steel and concrete are non-combustible materials and cannot cause integrity failure within the element. Integrity failure is usually caused by inability of the joints between construction elements to accommodate large deformations. After the Cardington fire tests (cf. Figure 3.33), this mode of failure is now emerging as an important design consideration. The principal measures to prevent this mode of integrity failure are to specify deflection limits and to provide continuity at the junctions.
- Insulation is concerned with thermal performance, which will usually be considered as part of load bearing calculations.

This book will only consider design calculations for steel and composite members. Readers should bear in mind that failure of a member in a complete structure does not necessarily lead to a structural collapse. Whether or not the loss of load carrying capacity of a structural member will lead to a collapse of the structure of which it forms part will depend on the nature of the structure, principally whether there is redundancy in the structure to provide alternative load paths. If, after failure of some structural members, reliable alternative load paths still exist, the structure will be stable. Nevertheless, once these load paths are identified, detailed design will be to ensure that each individual structural member in the critical load path has sufficient load carrying capacity.

Chapter 5 has shown that the behaviour history (displacements, stresses and strains) of a steel-framed structure in fire is complex. To simulate this complicated behaviour, sophisticated numerical models should be used. Design calculations are usually concerned with the ultimate strength of a structure in fire and use relatively simple hand calculation methods. Nevertheless, as mentioned earlier, deflections can become an important issue if large deflections lead to integrity failure. Unfortunately, no simple method is yet available to trace the displacement behaviour of a structure in fire.

This chapter will introduce design methods in both Part 8 and EC3 for calculating the ultimate load bearing capacity of structural members at the fire limit state. Wherever possible, a direct comparison between these two methods will be provided.

8.1.2   Loading

Fire is an accidental event and its coincidence with extreme structural loading is rare. Therefore, for structural FR design, reduced partial safety factors for structural loads should be used. Both Part 8 and Eurocode 1 Part 1.2 (CEN 2000a) give values of the reduced load factors for the fire limit state. These values are not repeated here but they are broadly similar to those used for serviceability design calculations at ambient temperature.

8.1.3   Performance of fire protection materials

It is assumed that a fire protection material is able to fulfill its intended functions during the life of the protected structure and also during a fire exposure. This may be ensured by making sure that the fire protection material can stick to the protected structure or by limiting deflections of the structure. Also fire protection materials may get damaged. At present, there are very few studies (Ryder *et al.* 2001; Tomecek and Milke 1993) of this problem and there is insufficient information to help develop a sensible simple design guide to assess the acceptable extent of damage to fire protection materials. In the light of this, the designer has to ensure that any damage to the fire protection is repaired.

8.1.4   Fire resistant design according to BS 5950 Part 8

The British Standard BS 5950 Part 8 is the first formal code of practice to allow the fire resistance of a steel structural element to be calculated. It is mainly based on findings of the standard fire resistance tests on simply supported individual steel structural members. Hence, there are a number of fundamental limitations to its applications, listed below:

1   Only the standard fire exposure can be dealt with.
2   There is no consideration of structural continuity and structural interactions in a complete building. This implies that loads and boundary conditions of a structural element remain unchanged during fire exposure.
3   It is only applicable to hot rolled sections, although, recently Lawson (1993) adapted Part 8 to the design of cold-formed thin-walled steel structural members in fire.

8.1.5   Fire resistant design according to Eurocode 3 Part 1.2

EC3 is part of a set of European design standards for fire safety of steel structures which include Eurocode 1 Part 1.2 (CEN 2000a) for fire loading calculations. Eurocode 1 Part 1.2 gives a method to evaluate the temperature–time relationships of realistic fires (cf. Section 7.3.6). Thus, EC3 is able to deal with more realistic fire conditions and has a wider scope of application than Part 8.

Unlike Part 8, EC3 gives some limited consideration of the effect of structural continuity on steel columns. Essentially, EC3 allows the column slenderness to be reduced if the column in fire meets certain requirements, thus improving the fire resistance of the column. This will be discussed in more detail in Section 8.4.2. At present, Eurocode 3 Part 1.2 does not consider cold-formed thin-walled steel structures.

8.2 OVERVIEW OF DESIGN METHODS

8.2.1 BS 5950 Part 8

In Part 8, calculations are carried out in the temperature domain. It is assumed that each steel member has a critical element. For example, the lower flange is the critical element of a steel beam. When the temperature of this critical element in fire has reached a limiting temperature, the load carrying capacity of the structural member has decreased to the level of the applied load in fire. Implicit within this assumption is that the temperature distribution in the structural member should follow a pre-defined distribution. For example, for a column, it is assumed that the temperature distribution should be uniform. For a steel beam, it is assumed that fire attack is from underneath.

The main design objective of Part 8 is to obtain the limiting temperature of the critical element of a structural member. If the design temperature, which is the maximum temperature that will be reached in fire in the critical element of the structural member, is lower than the limiting temperature, the structural member has sufficient load carrying capacity and it is safe. Otherwise, the structural member may be unsafe and design should be modified. Part 8 gives design temperatures for unprotected steelwork and Figure 8.1 shows the design temperatures of unprotected steelwork at 30 and 60 min of standard fire rating. If the limiting temperature of the structural member is lower than those in Figure 8.1, fire protection may be necessary and Part 8 gives a direct method to calculate the required fire protection thickness according to the limiting temperature and the design temperature of the unprotected steelwork.

The limiting temperature of a steel structural member is a function of the applied load. In Part 8, load is represented by a load ratio, defined as the ratio of the applied load on the structural member in fire to its load carrying capacity at ambient temperature. The limiting temperature concept is illustrated in Figure 8.2. The design flow chart according to Part 8 is shown in Figure 8.3.

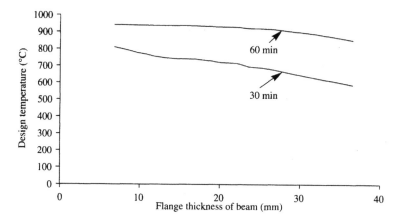

Figure 8.1 Temperatures in unprotected steel beams (from BSI 1990b). Reproduced with the permission of the British Standards Institution under licence number 2001SK/0298.

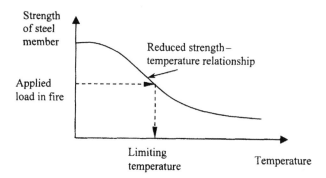

Figure 8.2 The limiting temperature method.

8.2.2   Eurocode 3 Part 1.2

EC3 adopts the design procedure at ambient temperature, but using the reduced strength and stiffness of steel at elevated temperatures in fire. It directly calculates the reduced load bearing capacity of a steel structural member and then compares it against the applied load in fire. Design is satisfactory when the reduced strength of the steel structural member is not less than the applied load in fire.

Using EC3, the fire exposure condition is not limited to the standard heating condition and more realistic fire exposures may be dealt with by

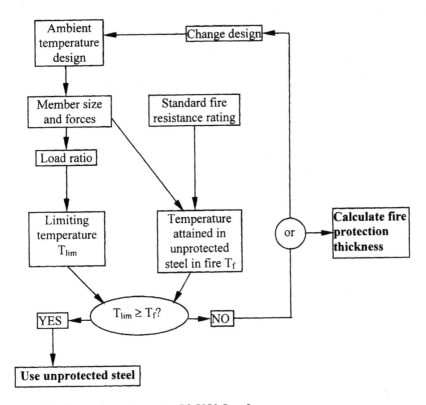

Figure 8.3 Design flow chart using BS 5950 Part 8.

using Eurocode 1 Part 1.2 (CEN 2000a). The flow chart of using EC3 is given in Figure 8.4.

Part 8 operates in the temperature domain and calculations are carried out in a sequential way, i.e. the FR design calculations follow the ambient temperature design calculations and no prior assumption is necessary. EC3 calculations are in the load domain. The EC3 method is conceptually more straightforward, however, it is an iterative process and design is carried out by trial and error.

8.3   FIRE RESISTANT DESIGN OF STEEL BEAMS

8.3.1   BS 5950 Part 8

For beams which are bending members, Part 8 gives two methods: (1) either the limiting temperature method; or (2) the moment capacity method may be used.

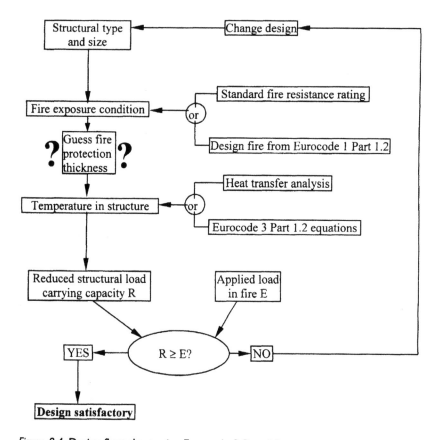

Figure 8.4 Design flow chart using Eurocode 3 Part 1.2.

8.3.1.1 Limiting temperature method

Load ratio

For a steel beam under bending, the load ratio (R) is defined as:

$$R = \frac{M_f}{M_c} \tag{8.1}$$

where M_f is the maximum bending moment in the beam under fire condition and M_c the bending moment capacity of the beam at ambient temperature. If the beam fails by complete yielding of steel, M_c is equal to the plastic bending moment capacity of the cross-section of the beam from equation (2.1). If LTB governs design, M_c is the LTB resistance of the beam from equation (2.3).

Table 8.1 Limiting temperatures of steel beams

Conditions of application	Limiting temperature (°C) at a load ratio of						
	0.7	0.6	0.5	0.4	0.3	0.2	0.1
Case 1	520	555	585	620	660	715	N/A
Case 2	460	510	545	590	635	690	N/A
Case 3	590	620	650	680	725	780	880
Case 4	540	585	625	655	700	745	811

Source: BSI (1990b). Reproduced with the permission of the British Standards Institution under licence number 2001SK/0298.

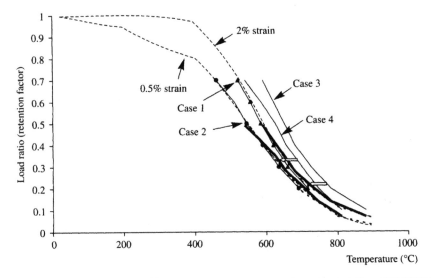

Figure 8.5 Limiting temperature–load ratio relationships of steel beams (from BSI 1990b). Reproduced with the permission of the British Standards Institution under licence number 2001SK/0298.

The limiting temperature of a beam is obtained from the limiting temperature–load ratio relationships in Table 8.1. Figure 8.5 compares the limiting temperature–load ratio relationships with the relationships of temperature–reduced strength of steel at different strain levels.

Four cases should be considered when applying this method:

Case 1:   This applies to a steel beam that is uniformly heated and can undergo large strains before failure. The beam is either unprotected or protected. If it is protected, the fire protection material should be able to undergo large strains without detaching from the protected beam. Here failure of the steel beam is essentially by steel yielding. Therefore, as shown in Figure 8.5, the limiting temperature–load ratio relationship coincides with the temperature–steel strength at 2% strain relationship.

Case 2:   This applies to a uniformly heated beam that is protected with a fire protection material that cannot sustain large strains. It is assumed that the fire protection material can safely sustain a strain of 0.5%. Thus, as shown in Figure 8.5, the limiting temperature–load ratio relationship follows the steel stress at 0.5% strain–temperature relationship.

Case 3:   This applies to a steel beam that is heated on three sides only with the top flange supporting and being protected by concrete slabs. Therefore, temperature distribution in the steel cross-section is non-uniform. Similar to a Case 1 beam, the steel beam is either unprotected or the fire protection material can undergo large strains before failure. Because the top flange is much cooler than the critical bottom flange, the beam has a higher bending moment capacity than that calculated assuming uniform temperature distribution when the lower flange temperature is the same.

Case 4:   The fire exposure condition is the same as in Case 3 but the steel beam is protected with a fire protection material that cannot undergo large strains.

Case 3 and Case 4 beams are related to Case 1 and Case 2 beams with a factor of 0.7 applied to the load ratio. The plastic bending moment capacity of a beam with uniform temperature distribution is approximately 0.7 times that with non-uniform distribution when the lower flange temperature is the same. The bold lines in Figure 8.5 are obtained by multiplying the load ratios of Case 3 and Case 4 beams by 0.7. It can be seen that the bold lines are very close to the Case 1 and Case 2 beams having uniform temperature distribution.

The limiting temperature method is based on steel beams that fail by complete yielding of steel at high temperatures. Because the limiting temperature method will also be used for steel beams that fail by LTB, the implicit assumption in Part 8 is that reductions in the strength and stiffness of steel at elevated temperatures are similar. The implication of this assumption will be further discussed in Section 8.3.3.

8.3.1.2 Bending moment capacity method

The limiting temperature method is simple to use. However, if temperature distribution in a steel beam is non-uniform and the beam is prevented from LTB, it may sometimes be beneficial to use a more accurate method to calculate the bending moment capacity of the beam. Such calculations may be carried out using the bending moment capacity method. In this method, the cross-section of the steel beam is divided into a number of thin slices of approximately the same temperature. The plastic bending moment capacity of the cross-section is calculated according to the reduced strengths of steel at the temperatures of these slices.

Table 8.2 An example of using the bending moment capacity method

Part i	Temperature T_i (°C)	Resultant force $F_i = p_y(T_i) \times A_i$(kN)	Lever arm from top d_i (mm)	Bending moment $= F_i \times d_i$ (kN.m)
Upper flange	$650 \times 0.8 = 520$	−565.81	8	−4.53
Web 1	528.13	−86.08	39.8	−3.43
Web 2	544.38	−24.10	70.76	−1.71
		55.96	94.56	5.29
Web 3	560.63	73.93	135.0	9.98
Web 4	576.88	67.74	182.6	12.37
Web 5	593.13	61.56	230.2	14.17
Web 6	609.38	55.75	277.8	15.49
Web 7	625.63	50.21	325.4	16.34
Web 8	641.88	44.68	373.0	16.66
Lower flange	650.0	266.16	404.8	107.74
Total	NA	0	NA	188.39

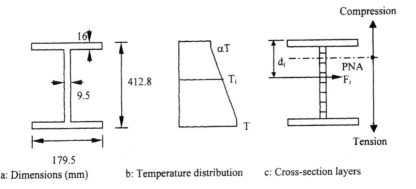

a: Dimensions (mm)   b: Temperature distribution   c: Cross-section layers

Figure 8.6 Input data for calculations in Table 8.2.

An example is given in Table 8.2 to illustrate this method. Input information for this example are shown in Figure 8.6. As can be seen, the bending moment capacity method requires many calculations. Hence, the only benefit of using this more elaborate method is to explore possible benefits of the non-uniform temperature distribution in the cross-section of the beam, particularly to justify the use of unprotected steelwork. Because of this, the steel stress at 2% strain should be used.

8.3.1.3 Comparison between the limiting temperature method and the bending moment capacity method

Given a universal beam size of 406 × 178 × 74 UB with dimensions in Figure 8.6, use the bending moment capacity method to calculate the plastic bending moment capacity of the cross-section in fire for load ratios of 0.3, 0.5 and 0.7. The lower flange temperature of the beam is equal to that obtained from the limiting temperature method. Consider different non-uniform temperature distributions as shown in Figure 8.6. Use S275 steel.

Example of solution for load ratio $R = 0.5$ and temperature gradient factor $\alpha = 0.8$

i   *Limiting temperature method*:   The plastic bending moment capacity of the cross-section at ambient temperature is $0.275 \times 1484.1 = 408.1$ kN.m. For a load ratio of 0.5, the plastic bending moment capacity of the beam in fire is 204.05 kN.m.

ii *Bending moment capacity method*: At a load ratio of 0.5, the limiting temperature of the beam is $650\,°C$ (Table 8.1, Case 3). This is the lower flange temperature. To apply the bending moment capacity method, divide the cross-section of the beam into ten parts as shown in Figure 8.6: the lower flange, the upper flange and eight equal parts of the web. Each part is assumed to have a constant temperature, calculated at the centroid. Results of these calculations are listed in Table 8.2. For each part of the cross-section, Table 8.2 gives the centroid temperature, the resultant force, the lever arm and the contribution to the plastic bending moment capacity of the cross-section. Since the PNA is in Web 2, this part is further divided into two layers: (1) the upper part in compression; and (2) the lower part in tension.

Using the same procedure, the plastic bending moment capacity of the cross-section for different combinations of load ratio and temperature gradient have been calculated and results are given in Table 8.3.

The three temperature gradients of $\alpha = 0.6$, 0.7 and 0.8 represent the realistic range of non-uniform temperature distribution in a steel beam exposed to fire on three sides with the upper flange protected by floor slabs. At the limiting temperature of the lower flange, the temperature difference is about $100–300\,°C$ between the upper and lower flanges.

Table 8.3 Results of using the plastic bending moment capacity

Load ratio	T_{lim} (°C)	Limiting temperature method	From the bending moment capacity method for a temperature gradient factor of			
			$\alpha = 0.6$	$\alpha = 0.7$	$\alpha = 0.8$	$\alpha = 1.0$
$R = 0.3$	725	122.43	145.34 ($\Delta T = 290\,°C$)	130.85 ($\Delta T = 218\,°C$)	116.74 ($\Delta T = 145\,°C$)	79.58 (uniform)
$R = 0.5$	650	204.05	217.18 ($\Delta T = 260\,°C$)	205.95 ($\Delta T = 195\,°C$)	188.39 ($\Delta T = 130\,°C$)	137.53 (uniform)
$R = 0.7$	590	285.67	282.63 ($\Delta T = 236\,°C$)	274.41 ($\Delta T = 177\,°C$)	260.75 ($\Delta T = 118\,°C$)	205.93 (uniform)

The following observations can be made from the results in Table 8.3:

- Under realistic non-uniform temperature distributions in a steel beam, the beam bending moment capacity is not very sensitive to the exact temperature in the upper flange of the beam.
- Results from the limiting temperature method are reasonably close to those from the bending moment capacity method for all load ratios.
- The bending moment capacity method may be used to give the beam an enhanced bending moment capacity if the temperature gradient is high so as to give a temperature difference between the lower flange and the upper flange of about 250 °C or higher.

8.3.2   Eurocode 3 Part 1.2 method

8.3.2.1   Plastic bending moment capacity

EC3 gives two methods to calculate the plastic bending moment capacity of a steel beam. The first method is identical to the bending moment capacity method in Part 8, which has been demonstrated in Table 8.2. Since this method requires many calculation steps, EC3 gives an alternative method which is much simpler to use. In this simple method, the plastic bending moment capacity of a beam is given by

$$M_{p,fi} = k_{y,T} M_p / (\kappa_1 \kappa_2) \tag{8.2}$$

where M_p is the bending moment capacity of the cross-section at ambient temperature; $k_{y,T}$ is the reduction factor in the effective yield strength of steel at the maximum temperature in the lower flange. Thus, $k_{y,T}M_p$ gives the reduced plastic bending moment capacity of the cross-section at a uniform temperature T. Adaptation factors κ_1 and κ_2 are used to account for non-uniform temperature distributions in the steel beam: κ_1 is for non-uniform temperature distribution in the cross-section; and κ_2 is for calculating the support hogging bending moment capacity when there is non-uniform temperature distribution along the beam. For the previous example, $\kappa_1 = 0.7$ and $\kappa_2 = 1.0$.

In Table 8.3, results are also given for the plastic bending moment capacity of the cross-section with uniform temperature distribution. Dividing these values by 0.7 gives $M_{p,fi}$ of 113.69 kN.m, 196.47 kN.m and 294.19 kN.m, respectively for the three load ratios. It can be seen that these values are very close to those in Table 8.3 from the bending moment capacity method. Hence, it may be concluded that for steel beams with realistic non-uniform temperature distributions, the limiting temperature method in Part 8, the bending moment capacity method in both codes and the simple alternative method in EC3 all give very similar results for the plastic bending moment capacity of the cross-section. However, the bending moment capacity method is the most laborious to implement. Therefore, unless there is a strong reason (e.g. to justify the elimination of fire protection) to use the bending moment capacity method, the limiting temperature method is preferred in Part 8 and the simple alternative method in EC3.

8.3.2.2 Lateral torsional buckling of steel beams

Although the bending resistance of a steel beam is usually controlled by the plastic bending moment capacity of the cross-section, it is sometimes necessary to consider the LTB resistance of a steel beam in fire.

In EC3, the LTB resistance of a steel beam is calculated in the same way as at ambient temperature. The calculation procedure is shown below.

The beam slenderness at elevated temperatures is given by

$$\overline{\lambda}_{LT,fi} = \overline{\lambda}_{LT} \sqrt{\frac{k_{y,T,com}}{k_{E,T,com}}} \qquad (8.3)$$

where $\overline{\lambda}_{LT}$ is the beam LTB slenderness at ambient temperature (calculated using Eurocode 3 Part 1.1 (CEN 1992a)); $k_{y,T,com}$ and $k_{E,T,com}$ are retention factors for the strength and elastic modulus of steel at the temperature of the compression flange. Since the behaviour of the compression flange is critical to the LTB of a beam, it appears reasonable to use the compression flange temperature to modify the beam slenderness at elevated temperatures. However, as will be shown later (Table 8.4), using equation (8.3) can lead to results that contradict realistic expectations.

Using the modified beam slenderness for fire, the strength reduction factor χ_{fi} for LTB in fire is calculated as:

$$\chi_{LT,fi} = \frac{1}{\phi_{LT,fi} + \sqrt{\left(\phi_{LT,fi}^2 - \overline{\lambda}_{LT,fi}^2\right)}},$$

$$\phi_{LT,fi} = 0.5\left[1 + \alpha_{LT}\left(\overline{\lambda}_{LT,fi} - 0.2\right) + \overline{\lambda}_{LT,fi}^2\right] \qquad (8.4)$$

where α_{LT} is a constant to account for the effect of initial imperfections and $\alpha_{LT} = 0.21$ for rolled sections.

The beam LTB resistance in fire ($M_{b,fi}$) is calculated using:

$$M_{b,fi} = \frac{\chi_{LT,fi}}{1.2} M_{p,fi} \qquad (8.5)$$

where $M_{p,fi}$ is the plastic bending moment resistance of the cross-section in fire, calculated using equation (8.2) or the bending moment capacity method.

Table 8.4 Comparison between Part 8 and EC3 calculations for lateral torsional buckling

Load Case Lateral torsional buckling resistance calculated using Part 8 and EC3 for a beam slenderness $\bar{\lambda}_{LT}$ at ambient temperature of

ratio	Load Case	0.208(20) BS	EC	0.411(40) BS	EC	0.606(60) BS	EC	0.790(80) BS	EC	0.962(100) BS	EC	1.342(150) BS	EC
0.7	(1)	0.7	0.716	0.698	0.556	0.606	0.507	0.5	0.435	0.412	0.354	0.265	0.213
	(2)		0.716		0.554		0.502		0.426		0.343		0.204
	(3)		0.716		0.553		0.498		0.418		0.333		0.197
	(4)	0.7	0.718	0.698	0.558	0.606	0.508	0.5	0.435	0.412	0.353	0.265	0.212
	(5)		0.718		0.554		0.499		0.420		0.334		0.197
	(6)	0.7	–	0.657	–	0.534	–	0.419	–	0.329	–	0.199	–
0.5	(1)	0.5	0.5	0.498	**0.386**	0.427	**0.348**	0.357	**0.293**	0.294	**0.234**	0.189	**0.138**
	(2)		0.5		0.389		0.355		0.306		0.249		0.150
	(3)		0.5		0.386		0.348		0.292		0.233		0.137
	(4)	0.5	0.517	0.498	0.398	0.427	0.358	0.357	0.300	0.294	0.238	0.189	0.140
	(5)		0.517		0.399		0.359		0.302		0.241		0.142
	(6)	0.5	–	0.469	–	0.381	–	0.299	–	0.235	–	0.142	–
0.3	(1)	0.2	0.191	0.199	0.149	0.171	0.135	0.143	0.116	0.118	0.094	0.076	0.56
	(2)		0.191		0.147		0.131		0.108		0.085		0.049
	(3)		0.191		0.148		0.133		0.112		0.089		0.053
	(4)	0.2	0.212	0.199	0.161	0.171	0.141	0.143	0.113	0.118	0.087	0.076	0.050
	(5)		0.212		0.164		0.147		0.124		0.099		0.058
	(6)	0.2	–	0.188	–	0.152	–	0.120	–	0.094	–	0.057	–

Note
The slenderness in brackets is according to BS 5950 Part 1 definition.

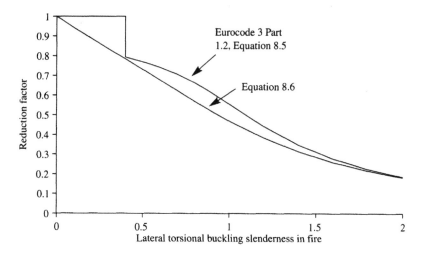

Figure 8.7 Strength reduction factor–slenderness relationship for lateral torsional buckling of a steel beam in fire.

The modification factor 1.2 in equation (8.5) is a correction factor for the fire situation. It is used to allow for a number of effects, e.g. the stress–strain curve of steel is no longer linear at elevated temperatures. LTB is not considered if the beam slenderness $\overline{\lambda}_{LT,fi}$ does not exceed 0.4. Thus as shown in Figure 8.7, a step change is introduced in the beam LTB resistance at a slenderness of $\overline{\lambda}_{LT,fi} = 0.4$.

It is noted from equations (8.2) and (8.3) that the lower flange temperature of a beam determines the plastic bending moment capacity of the cross-section and the upper flange temperature determines the beam slenderness. Thus, if the lower and upper flange temperatures of the steel beam are given, calculations for the LTB resistance of the beam are straightforward. However, in realistic situations, load is given and it is required to find the temperatures of the beam at failure. Under this condition, an iterative procedure has to be used. In this iterative process, the steel temperatures are guessed and the LTB resistance of the beam is calculated. This is then compared to the applied load in fire to determine appropriate temperature values for the next iteration until the temperature difference between two successive iterations become acceptably small.

8.3.3   A comparison between Part 8 and EC3 for lateral torsional buckling

Compared with the iterative process for calculating the beam failure temperature in EC3, the limiting temperature approach in Part 8 is much simpler to use. It is thus interesting to compare predictions of these two methods. The results of such a comparison are given in Table 8.4. Uniform bending along the beam span is assumed. This table gives the calculated LTB resistance of a steel beam with a cross-section $406 \times 178 \times 74\,UB$. Variables considered in this comparison are the beam LTB slenderness, the beam lower flange temperature and the beam upper flange temperature. The three beam lower flange temperatures are obtained using the limiting temperature method of Part 8 at load ratios of 0.7, 0.5 and 0.3. The upper flange temperature is lower than the lower flange temperature by 0 °C, 150 °C and 250 °C. A difference of 0 °C gives uniform heating in the beam. Differences of 150 °C and 250 °C are the two likely extremes of non-uniform temperature distribution in the cross-section of a beam.

Table 8.4 gives ratios of the LTB bending moment resistance of a beam at elevated temperatures to the plastic bending moment resistance of the cross-section at ambient temperature. For each load ratio, the following six cases have been considered:

Case 1: The upper flange temperature is lower than the lower flange temperature by 250 °C and the beam slenderness for EC3 calculations is obtained from equation (8.3).

Case 2: The upper flange temperature is lower than the lower flange temperature by 150 °C and the beam slenderness for EC3 calculations is obtained from equation (8.3).

Case 3: The beam is non-uniformly heated and the beam slenderness is obtained by multiplying the ambient temperature value by 1.2 instead of using equation (8.3).

Case 4: The beam is uniformly heated and the beam slenderness for EC3 calculations is obtained from equation (8.3).

Case 5: The beam is uniformly heated and the beam slenderness is obtained by multiplying the ambient temperature value by 1.2 instead of using equation (8.3).

Case 6: Calculations follow Part 8 equations except that the beam slenderness in fire is obtained by multiplying the beam slenderness at ambient temperature by 1.2.

In Cases 1–5, the beam slenderness for Part 8 calculations is unchanged. Since Part 8 calculations do not depend on temperatures and the beam slenderness is unchanged, the same value is obtained for Cases 1–5 from Part 8 calculations.

From the results presented in Table 8.4 a number of observations can be made:

- At a given lower flange temperature, the results from EC3 (Cases 1 and 2) are not sensitive to temperatures in the upper (compression) flange of the beam, except for very high beam slenderness and low load ratio.
- In EC3 calculations, if the beam slenderness is modified by using equation (8.3), it is possible that at the same lower flange temperature,

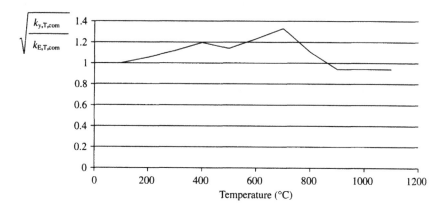

Figure 8.8 Modification factor for the lateral torsional buckling slenderness of a beam in fire (equation (8.3)).

a higher top flange temperature can lead to an increased beam LTB resistance (bold numbers in Table 8.4). This is contradictory to expectation and is purely due to the fact that only the upper (compression) flange temperature is used to modify the beam LTB slenderness. As shown in Figure 8.8, this modification factor does not increase monotonically with increasing temperature. At temperatures between 400 °C and 500 °C and greater than 700 °C, this modification factor decreases with increasing temperatures. Thus, if the upper flange temperature is within the above regions, a higher upper flange temperature will give a reduced beam slenderness using equation (8.3). If the lower flange temperature of the beam is the same and hence the plastic bending moment capacity is the same, the LTB resistance of the beam is higher at a higher temperature in the upper flange.

- Figure 8.8 shows that at temperatures in the realistic region of 300 °C–800 °C, the slenderness modification factor in equation (8.3) is very close to a constant of 1.2. Using this constant instead of the modification factor in equation (8.3), there is very little difference in the final results by comparing results of Case 3 to those of Cases 1 and 2 for non-uniform heating and results of Case 5 to those of Case 4 for uniform heating. However, the benefit of using a constant of 1.2 is a great simplification of the calculation process and no iteration is necessary.

- At a low beam slenderness ($\overline{\lambda}_{LT} = 0.208$ or $\lambda_{LT} = 20$) where plastic yielding of the cross-section governs failure, both Part 8 and EC3 give very similar results. For more slender beams where LTB becomes more important, Part 8 gives a higher resistance than EC3. In addition to the difference at ambient temperature, two factors contribute to the difference between these two methods in fire: (1) In Part 8, the beam slenderness is not increased at elevated temperatures; (2) A reduction factor of 1.2 is applied on the resistance side in EC3 in equation (8.5). By coincidence, for the beam studied in this example, at ambient temperature and high slenderness ($\bar{\lambda}_{LT} > 0.79$ or $\lambda_{LT} > 80$), the LTB resistance of the beam calculated using Eurocode Part 1.1 is about 1.2 times that calculated using BS 5950 Part 1 (BSI 1990a, cf. Figure 2.7). This cancels out the effect of using a reduction factor of 1.2 in equation (8.5). Therefore, using a different slenderness becomes the only reason for differences in the calculated results between Part 8 and EC3. When the same beam slenderness is used in both Part 8 and EC3 calculations (compare results of Case 6 from Part 8 with those of Cases 1–5 from EC3 in Table 8.4), both methods give very similar results.

From this comparative study, it can be concluded that Part 8 and EC3 produce similar results for the LTB resistance of a beam in fire. The difference in the predicted results using the two methods is solely because they use different values of LTB slenderness for the beam at elevated temperatures. It seems reasonable that the LTB slenderness in Part 8 should be modified to reflect different changes in the strength and stiffness of steel at elevated temperatures. As an approximation, the beam slenderness in fire is 1.2 times that at ambient temperature.

8.3.4   Modifications to the EC3 method

In both Part 8 and EC3 calculations, the ambient temperature design method is retained. The implied assumption is that the steel stress–strain relationship is linear elastic–plastic. However, at elevated temperatures, the stress–strain relationship of steel is highly non-linear and depending on the applied stress level, the tangent stiffness of the beam can be much less than the initial value. Consider dividing the slenderness of the beam into three regions. With a high beam slenderness, the applied stress is generally low and the beam tangent stiffness is close to the initial value. For a beam with a low slenderness where steel yield governs design, the beam bending resistance is not sensitive to its tangent stiffness. However, for a beam with a medium slenderness ($\bar{\lambda}_{LT,fi} \approx 1$) where the applied stress is substantial and LTB is important, the ambient temperature design approach can lead to gross overestimation of the LTB resistance of the beam in fire. Although EC3 introduces a constant of 1.2 in equation (8.5) to account for this, this broad brush approach cannot simulate the complicated behaviour of LTB of a steel beam in fire. LTB of steel beams in fire has been studied by Bailey et al. (1996a) and Vila Real and Franssen (2000), both using finite element techniques. Both investigators have come to the same conclusion.

As a modification to the approach in EC3, Vila Real and Franssen (2000) have suggested an alternative method to calculate the strength reduction factor for LTB. Instead of using equation (8.5), the modified strength reduction factor for LTB resistance is given by

$$\chi_{LT,fi} = \frac{1}{\phi_{LT,fi} + \sqrt{\phi^2_{LT,fi} - \overline{\lambda}^2_{LT,fi}}} \quad \text{with}$$

$$\phi_{LT,fi} = \frac{1}{2}\left[1 + \alpha\overline{\lambda}_{LT,fi} + \left(\overline{\lambda}_{LT,fi}\right)^2\right] \tag{8.6}$$

In equation (8.6), $\alpha = \beta\varepsilon$ is an imperfection factor and $\beta(=0.65)$ is the so-called severity factor, obtained from curve-fitting to the numerical results predicted using the computer program SAFIR, $\varepsilon = \sqrt{235/p_y}$. Figure 8.7 compares the strength reduction factors obtained using equations (8.5) and (8.6). As intended, the approach of Vila Real and Franssen gives a lower LTB resistance for beams of medium slenderness.

The proposed modification by Vila Real and Franssen has two advantages:

1   It eliminates the overestimation in EC3;
2   It eliminates the step change at $\overline{\lambda}_{LT,fi} = 0.4$.

8.3.5   Temperatures in steel beams

To perform calculations as outlined in previous equations, it is necessary to have available information on temperatures in a steel beam. For an unprotected steel beam or a protected steel beam with approximately uniform temperature, equations (6.40) and (6.45) may be used. For a number of special steel beams which are integrated with concrete, a number of empirical methods have been developed for the standard fire exposure and these methods are presented in the following paragraphs. Under realistic fire conditions, numerical methods should be used.

8.3.5.1   Slim floor beams

Fellinger and Twilt (1996) have devised a simple method to evaluate temperatures in different parts of the steel section of a slim floor beam. Their method is based on the results of a comprehensive numerical study using the general finite element program DIANA. In their method, the assumed temperature distribution is as shown in Figure 8.9.

For temperatures in the lower flange and the plate, equation (8.7) may be used:

$$T = C_1 - C_2 t_f - C_3 t_p \tag{8.7}$$

where t_f and t_p are the flange and plate thickness in mm respectively; C_1, C_2 and C_3 are coefficients depending on the standard fire resistance rating, the

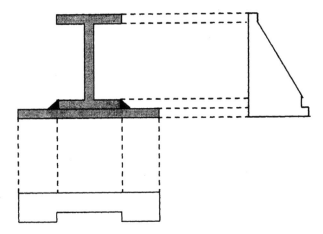

Figure 8.9 Temperature distributions in a slim floor beam assumed by Fellinger and Twilt (1996).

location of the steel component and whether or not there is an air gap between the steel plate and the lower flange of the steel section caused by welding. Values of C_1 and C_2 and C_3 are given in Table 8.5.

Temperatures in the upper flange are not considered since they are considerably lower than 400 °C at which reduction in steel strength starts. Temperatures in the web are assumed to have a linear distribution. To

Table 8.5 Coefficients for using equation (8.7)

Fire rating (min)	Location of steel	Without air gap			With an air gap of 4 mm		
		C_1 (°C)	C_2 (°C/mm)	C_3 (°C/mm)	C_1 (°C)	C_2 (°C/mm)	C_3 (°C/mm)
30	plate	570	5.0	5.0	730	1.5	8.0
	plate tip	675	2.0	7.5	700	1.5	7.0
	lw. flange	550	3.0	4.4	450	2.0	5.8
60	plate	850	3.0	3.0	850	1.5	2.0
	plate tip	920	2.0	3.5	900	2.0	2.0
	lw. flange	850	3.8	3.5	750	3.5	3.8
90	plate	930	1.0	1.0	930	1.0	1.0
	plate tip	980	0.5	2.0	970	1.0	1.0
	lw. flange	925	1.8	1.8	840	2.0	1.8
120	plate	980	0	0	980	0	0
	plate tip	1010	0	0	1010	0	0
	lw. flange	980	1.4	1.4	920	1.5	1.7

Source: Fellinger and Twilt (1996).

Table 8.6 Comparison of lower flange temperatures in an ASB

	Lower flange temperature (°C) at fire resistance time of		
	30 min	60 min	90 min
Calculated (Fellinger and Twilt 1996)	496	782	893
Test (Ma and Makelainen 1999b)	500	760	890

obtain the web temperatures, the top flange temperature is assumed to be linearly related to the standard fire exposure time according to:

$$T_{uf} = 1.5t + 20 \; (°C) \tag{8.8}$$

where the standard fire resistance time t is in minutes.

Fellinger and Twilt (1996) compared the calculated plastic bending moment capacity of the beam using the approximate temperatures and using temperatures obtained from finite element analyses. They found that the difference in the plastic bending moment capacity of the steel cross-section was at most 5%.

8.3.5.2 Asymmetrical beams (ASB)

The method of Fellinger and Twilt (1996) described above may also be used to give approximate temperatures in different parts of an asymmetrical beam. To use this method, the plate thickness should be taken as 0 and there is no air gap. Also the lower flange temperature distribution is uniform. Table 8.6 compares the calculated results using the modified Fellinger and Twilt method and the measured temperatures by Ma and Makelainen (1999b). The accuracy of the modified Fellinger and Twilt method is acceptable.

8.3.5.3 Shelf angle beams

For shelf angle beams under the standard fire exposure, Part 8 gives a tabulated method to evaluate the temperatures in various parts of the steel section at a number of standard fire ratings.

8.3.6 Summary of fire resistant design calculations for lateral torsional buckling resistance

The EC3 approach involves iteration and is much more complicated to implement than the Part 8 approach. However, it is felt that the basis of the

EC3 approach is correct. To help implement the Eurocode 3 approach, the beam slenderness modification factor in equation (8.3) may be replaced by a constant 1.2. If this is adopted, calculations for the failure temperature of a beam in LTB does not require iteration and the steps involved are listed below:

- The LTB slenderness of a beam at elevated temperatures is calculated from $\bar{\lambda}_{LT,fi} = 1.2\bar{\lambda}_{LT}$, where the beam slenderness $\bar{\lambda}_{LT}$ will have already been calculated at ambient temperature. Either equation (8.4) or equation (8.6) may be used to calculate the strength reduction factor $\chi_{LT,fi}$ for lateral torsional buckling in fire.
- Given the applied bending moment in fire (M_{fi}), the required plastic bending moment capacity of the cross-section of the beam in fire is calculated from $M_{p,fi} = 1.2M_{fi}/\chi_{LT,fi}$ if the EC3 approach is used (equation 8.5), or from $M_{p,fi} = M_{fi}/\chi_{LT,fi}$ if the approach of Vila Real and Franssen (equation 8.6) is used.
- Using equation (8.2), the required steel strength retention factor $k_{y,T}$ is calculated from $k_{y,T} = \kappa_1 \kappa_2 M_{p,fi}/M_p$, where M_p is the plastic bending moment capacity of the cross-section at ambient temperature.
- The failure temperature of the beam is obtained from Table 5.2 in Chapter 5 to give a value of $k_{y,T}$ calculated above.

8.4   FIRE RESISTANT DESIGN OF STEEL COLUMNS

8.4.1   Axially loaded columns with uniform temperature distribution

8.4.1.1   BS 5950 Part 8

The limiting temperature method is used in Part 8. For an axially loaded column, the load ratio is defined as:

$$R = \frac{P_{fi}}{A_g p_c} \tag{8.9}$$

where P_{fi} is the applied axial load in fire, A_g the gross cross-sectional area and p_c the column flexural buckling strength at ambient temperature. When calculating p_c the column slenderness is assumed to be the same as at ambient temperature.

The relationships between the column limiting temperature and the load ratio are given in Table 8.7. In Part 8, two sets of limiting temperatures are given, one for column slenderness not exceeding 70 and one for column slenderness exceeding 70. In the former case, column failure is assumed to occur at a steel strain of about 1.5%. Thus, the limiting temperature–load

Table 8.7 Limiting temperatures of columns

Column slenderness	Limiting temperature (°C) at a load ratio of					
	0.7	0.6	0.5	0.4	0.3	0.2
≤70	510	540	580	615	655	710
>70 but ≤180	460	510	545	590	635	635

Source: BSI (1990b). Reproduced with the permission of the British
Standards Institution under licence number 2001SK/0298.

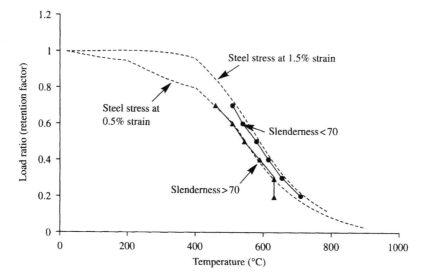

Figure 8.10 Limiting temperature–load ratio relationships of steel columns (from BSI
1990b). Reproduced with the permission of the British Standards Institution under licence number 2001SK/0298.

ratio relationship is very close to the steel stress at the 1.5% strain–temperature relationship, as shown in Figure 8.10. In the latter case, since the
column is slender, it is assumed that column failure occurs at a lower steel
strain of about 0.5% and the limiting temperature–load ratio relationship
follows the steel stress at the 0.5% strain–temperature relationship.

The effect of slenderness on the column limiting temperature has been
studied by a number of investigators (Najjar and Burgess 1996; Wang *et al.*
1995). Figure 8.11 shows the results of a comparison between the Part 8
limiting temperatures and the predicted column failure temperatures using a
finite element analysis by Wang *et al.* (1995). It can be seen that the Part 8
limiting temperature values are accurate for columns of low slenderness
($\lambda \leq 20$). For columns of high slenderness ($\lambda \geq 100$), the accuracy of the

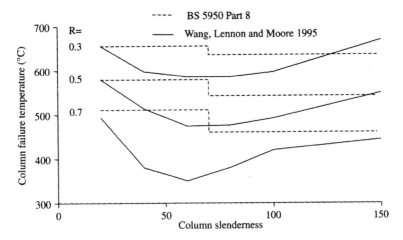

Figure 8.11 A comparative study of the effect of column slenderness on column failure temperature for different load ratios (from Wang *et al.* 1995).

Part 8 limiting temperature method is also reasonable. However, for columns of more realistic slenderness ($40 < \lambda < 100$), the Part 8 limiting temperatures are much higher than those calculated.

The overestimation of the column limiting temperature by Part 8 is due to the highly non-linear behaviour of stress–strain relationships of steel at elevated temperatures. At a low value of column slenderness ($\lambda \leq 20$), column failure is strength controlled and the Part 8 limiting temperature is accurate. If the column slenderness is high ($\lambda > 100$), the column buckling stress is low and the stress–strain relationship of steel at elevated temperatures is essentially linear. It is when the column slenderness is within the realistic range of 40–100 that discrepancy occurs. Here the steel stress is high and column failure is governed by elasto-plastic buckling where it is important to consider the effect of reduced tangent stiffness of the column.

Consider columns of different lengths giving slendernesses of 20, 60 and 100. According to Table 27(b) of BS 5950 Part 1 (BSI 1990a) at ambient temperature, the compressive strengths of such columns for S275 steel are 272, 221 and 74 N/mm², respectively. At a column load ratio of 0.7, the applied stresses are 190.4, 154.7 and 51.8 N/mm² for the three values of column slenderness. Notice the relatively small change in the applied steel stress from a column slenderness of $\lambda = 20$ to $\lambda = 60$. At these applied stresses, the required steel tangent modulus (E_{req}) for a stable column may approximately be calculated using the following equation:

$$E_{req} = \frac{\sigma \lambda^2}{\pi^2} \tag{8.10}$$

giving $E_{req} = 0.04$, 0.28 and 0.59 times the Young's modulus of steel at ambient temperature respectively. From the measured stress–strain curves of steel at elevated temperatures (Wainman and Kirby 1987), the column failure temperatures are 480 °C, 335 °C and 470 °C, respectively. While the failure temperatures of 480 °C and 470 °C for $\lambda = 20$ and $\lambda = 150$ are close to 510 °C and 460 °C given by Part 8 (Table 8.7), the failure temperature of 335 °C at $\lambda = 60$ is much lower than the Part 8 limiting temperature of 510 °C.

Thus, the column limiting temperature is not a fixed value for a given load ratio, but varies as a function of the column slenderness. To realistically predict the column failure (limiting) temperature, the slenderness effect should be properly considered.

8.4.1.2   Eurocode 3 Part 1.2

In EC3, calculations for the strength of a column follow the procedure in Eurocode 3 Part 1.1 (CEN 1992a), but modifications are made to account for high temperature effects. These modifications are: (1) the column slenderness is changed to allow for different rates of degradation in the strength and stiffness of steel; (2) A reduction factor of 1.2 is introduced to reduce the column buckling load in fire to account for the effects of non-linear stress–strain relationships of steel at elevated temperatures. These modifications are identical to those for LTB of a beam in fire.

The column slenderness is calculated using:

$$\bar{\lambda}_{fi} = \bar{\lambda}\sqrt{\frac{k_{y,T}}{k_{E,T}}} \tag{8.11}$$

where $\bar{\lambda}$ is the column slenderness at ambient temperature. $\bar{\lambda}$ is defined as:

$$\bar{\lambda} = \lambda\sqrt{\frac{p_y}{\pi^2 E}}, \quad \text{with} \quad \lambda = \frac{L_e}{r_y} \tag{8.12}$$

where L_e and r_y are the buckling length of the column and radius of gyration of the cross-section about the relevant axis of buckling.

The strength reduction factor is calculated using:

$$\chi_{fi} = \frac{1}{\phi_{fi} + \sqrt{\phi_{fi}^2 - \bar{\lambda}_{fi}^2}} \quad \text{with} \quad \phi_{fi} = 0.5\left[1 + \alpha(\bar{\lambda}_{fi} - 0.2) + \bar{\lambda}_{fi}^2\right] \tag{8.13}$$

where α is an imperfection factor.

Although there are four column buckling curves for column design at ambient temperature, for fire safety design in EC3, only the column curve "c" is used regardless of the type of cross-section and axis of buckling of the column, giving $\alpha = 0.49$. As can be seen in Figure 2.10 of Chapter 2, the column buckling curve "c" gives a relatively low value of column buckling resistance. Reasons for using such a low buckling curve in fire include more severe influences of initial imperfections and additional bending moments due to possible non-uniform temperature distributions.

The column resistance to axial compression is given by

$$P_{c,fi} = \frac{\chi_{fi}}{1.2} P_{u,fi} \tag{8.14}$$

where $P_{u,fi}$ is the column squash load at elevated temperature T and is calculated from:

$$P_{u,fi} = A_s k_{y,T} p_y \tag{8.15}$$

An iterative process is thus required to find the column failure temperature if the applied load in the column is given. However, by taking the column slenderness in fire as 1.2 times that at ambient temperature, as in the previous section for LTB of a beam, the design complexity is greatly reduced.

8.4.1.3 Comparison between BS 5950 Part 8 and Eurocode 3 Part 1.2

Comparing the calculation methods for LTB of a beam and flexural buckling of a column in both methods, it can be seen that they are identical in both Part 8 and EC3. It follows that the same conclusions may be drawn with regard to the accuracy of these design methods. To check whether this is the case, a comparative study has been performed.

The results of this comparative study are given in Table 8.8 where a range of column slenderness and limiting temperatures are considered. For each combination of column slenderness and column limiting temperature, Table 8.8 gives four ratios of the column buckling resistance in fire to the column squash load at ambient temperature, obtained using the following four methods:

1 BS 5950 Part 8;
2 BS 5950 Part 8, but the column slenderness in fire is 1.2 times that at ambient temperature;
3 Eurocode 3 Part 1.2;
4 Eurocode 3 Part 1.2, but the column slenderness in fire is 1.2 times that at ambient temperatures.

Table 8.8 Comparison between Part 8 and EC3 calculations for column design

T (°C)	Case	Buckling resistance of columns at an ambient temperature slenderness of					
		$\lambda = 20$ ($\bar{\lambda} = 0.233$)	40 (0.466)	60 (0.7)	80 (0.933)	100 (1.166)	150 (1.749)
510	(1)	0.687	0.606	0.512	0.352	0.273	0.149
	(2)	0.673	0.570	0.451	0.288	0.212	0.108
	(3)	0.603	0.514	0.413	0.313	0.234	0.122
	(4)	0.599	0.505	0.398	0.296	0.218	0.113
600	(1)	0.436	0.384	0.324	0.222	0.172	0.094
	(2)	0.426	0.361	0.285	0.182	0.133	0.069
	(3)	0.374	0.314	0.244	0.180	0.132	0.068
	(4)	0.376	0.317	0.250	0.186	0.137	0.071
700	(1)	0.214	0.189	0.159	–	–	–
	(2)	0.209	0.178	0.140	–	–	–
	(3)	0.181	0.148	0.111	0.080	0.057	0.029
	(4)	0.184	0.155	0.122	0.091	0.067	0.035

From the results in Table 8.8, the following observations may be made:

• If the second term in equation (8.11) is replaced by a constant of 1.2 (compare Case 3 and Case 4 results), there is very little loss of accuracy in the predicted column buckling resistance. However, the calculation procedure is greatly simplified.
• The constant of 1.2 in equation (8.14) is intended to account for the fact that the steel strain at column failure in fire is less than 2% (which is used in equation (8.15) to calculate the effective yield strength of steel at elevated temperatures). A comparison in Figure (8.12) shows that the effective yield strength of steel at 2% strain in EC3 is approximately 1.2 times the steel stress at 0.5% strain used in Part 8 for $\lambda > 70$. Hence, this constant of 1.2 in equation (8.15) is similar to using the reduced limiting temperature in Part 8 for $\lambda > 70$. This means that if both Part 8 and EC3 use the same column slenderness at elevated temperatures, the predicted column flexural buckling resistance in fire should be similar. Comparisons of results in Table 8.8 for Case 2 and Case 4 show this is indeed the case for $\lambda > 80$. For shorter columns ($\lambda \leq 70$) where failure is by plastic yield, the EC3 method gives lower column buckling resistance due to the universal application of the reduction factor of 1.2 in equation (8.14).

8.4.1.4 Proposed modification to EC3 by Franssen et al. (1995)

As for LTB of a steel beam, EC3 may overestimate the column buckling strength due to its inadequate consideration of the non-linear stress–strain

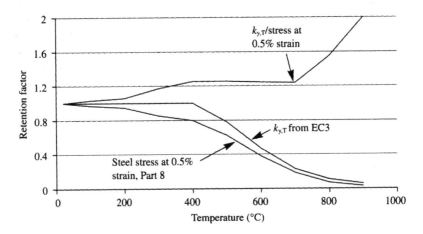

Figure 8.12 A comparison of retention factors used for column design.

relationship of steel at elevated temperatures. Again, the SAFIR program has been used to derive modifications to the EC3 design approach.

In the modified approach, the column strength reduction factor is given by

$$\chi_{fi} = \frac{1}{\phi_{fi} + \sqrt{\phi_{fi}^2 - \overline{\lambda}_{fi}^2}}, \quad \text{with} \quad \phi_{fi} = \frac{1}{2}\left(1 + \alpha\overline{\lambda}_{fi} + \overline{\lambda}_{fi}^2\right) \tag{8.16}$$

where α is an imperfection factor and $\alpha = \beta\varepsilon$. A value of $\beta = 0.65$ is recommended.

A comparison between the EC3 method and the proposed method of Franssen *et al.* is shown in Figure 8.13. It seems that in this case, the proposed method of Franssen *et al.* gives a slightly higher column strength than the EC3 method, implying that the combined effect of using a low column buckling curve (curve "c") and a reduction factor of 1.2 in equation (8.14) has eliminated the overestimation in EC3 which arises due to inadequate consideration of the non-linear stress–strain relationship of steel at elevated temperatures. Nevertheless, the method of Franssen *et al.* does give more logical results for columns at very low slenderness when steel yielding governs whilst the EC3 method gives a constant value of $0.83(=1/1.2)$.

8.4.1.5 Summary of column design under axial compression

In Part 8, the column slenderness for fire design should be modified to take into consideration different rates of reduction in the strength and stiffness of steel at high temperatures. Nevertheless, it is not necessary to use equation

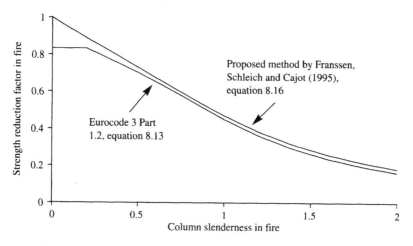

Figure 8.13 Comparison between column buckling loads.

(8.11). Instead the column slenderness in fire may be taken as 1.2 times that at ambient temperature.

The column design method in EC3 is conservative and can be used safely. To simplify its implementation, the column slenderness modification factor in equation (8.11) may be replaced by a constant of 1.2. The failure temperature of a column may be estimated in the following way:

- The column slenderness at elevated temperature is calculated from $\overline{\lambda}_{fi} = 1.2\overline{\lambda}$, where the column slenderness $\overline{\lambda}$ will already have been calculated at ambient temperature. Equation (8.13) is then used to calculate the column strength reduction factor χ_{fi}.
- Given the applied axial load in fire (P_{fi}), the required squash resistance of the column in fire is calculated from $P_{u,fi} = 1.2P_{fi}/\chi_{fi}$.
- The required steel strength retention factor $k_{y,T}$ is calculated using $k_{y,T} = P_{u,fi}/P_{u,a}$, where $P_{u,a}$ is the column squash load at ambient temperature.
- The failure temperature of the column is obtained from Table 5.2 in Chapter 5 to give a value of $k_{y,T}$ calculated from above.

8.4.2 Effects of structural continuity

8.4.2.1 Continuous construction

In discussions so far, the column is treated as an isolated member with no interaction with other structural members in a frame. This is rarely the case in a realistic structure and Section 5.3 has discussed possible ways that a column may interact with other members of a complete structure. They include:

- Changes in the column axial load due to restrained thermal movement;
- Changes in the column bending moments and effective buckling length due to changes in the column bending stiffness relative to the adjacent structure;
- Changes in the column bending moments due to lateral movement of the adjacent beams.

The only possible way to consider all these effects is by using advanced numerical analysis packages. To consider these effects in a suitable way for practical design, simplifications have to be made. From the discussions in Section 5.3, it has been concluded that the effects of the above-mentioned structural interactions on the behaviour of a column are as follows:

1   Restraining the axial expansion of a column generally increases the column axial compression which leads to earlier buckling of the column.
2   In a fire situation, a column loses more stiffness than does the adjacent structure. Therefore, the column end condition approaches fixity and the moments transferred from the connected beams to the column almost disappear.
3   There is a very small net change in the column bending moments due to thermal expansion of the adjacent beams.

Therefore, for practical design purposes, bending moments in a column in fire may be neglected. For axial behaviour, an increase in the column axial compression is detrimental to the column fire resistance but a reduction in the column effective length is beneficial. Neglecting the increase in the column axial compression will overestimate the column failure temperature and neglecting the reduction in the column buckling length will underestimate the column failure temperature. The net result of neglecting both effects will depend on the stiffness of the adjacent structure relative to that of the column. For a realistic steel framed building with a relative beam to column stiffness of about 2%, these two effects almost cancel out each other (Wang 1997b). For design simplicity, the column may be treated as pin ended but using the initial axial load regardless of the initial loading and boundary conditions.

Further complications arise with regard to the post-buckling behaviour of the column and different definitions of column failure. If these effects need to be considered, for example due to high restraint stiffness of the adjacent structure, the method in Section 5.3.2 may be used.

8.4.2.2 Simple construction

In simple construction with ideally pinned beam to column connections, there is no restraint to the column's thermal expansion but the beneficial

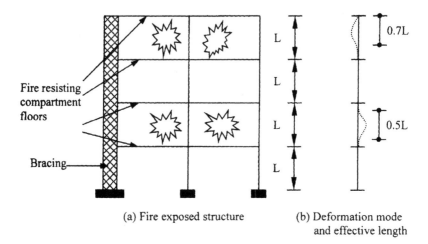

(a) Fire exposed structure   (b) Deformation mode
                                        and effective length

Figure 8.14 Column effective length for fire design (from CEN 2000b). Reproduced with the permission of the British Standards Institution under licence number 2001SK/0298.

effect of enhanced column support condition, which derives from restraints from the adjacent cold columns, is still present. Under this circumstance, the following recommendation in EC3 may be adopted:

> in the case of a steel frame in which each storey comprises a separate fire compartment with sufficient fire resistance, in an intermediate storey the buckling length of a column $l_{fi} = 0.5L$ and in the top storey the buckling length $l_{fi} = 0.7L$, where L is the system length in the relevant storey.

This statement is illustrated in Figure 8.14.

8.4.2.3   Summary of design considerations for structural continuity

The behaviour of a restrained column is complex. However, for design purposes, the following simplifications may be adopted:

● Column bending moments may be neglected;
● Increases in the column axial load due to restrained thermal expansion may be neglected;
● In continuous construction, the column effective length should be taken as its physical length, i.e. the effective length ratio is 1.0.
● In simple construction, the EC3 approach (illustrated in Figure 8.14) may be used to calculate the column effective length.

8.4.3   Columns with bending moments

The predominant loading on a column is axial compression, however, bending moments may also exist. Sources of bending moments in a column include lateral loads (e.g. wind load), eccentricity and if exact evaluation is possible, bending moments transferred from the adjacent structure in continuous construction.

The behaviour of a column under combined axial compression and bending moments is much more complicated than that under pure axial compression, but the problem has not been studied as intensively as that of simple axial load. As a result, design methods for this loading case are very approximate and further research is clearly needed.

8.4.3.1   BS 5950 Part 8

The limiting temperature method is used and the load ratio is obtained from the maximum of the following two equations
For local capacity:

$$R = \frac{P_{fi}}{A_g p_y} + \frac{M_{x,fi}}{M_{cx}} + \frac{M_{y,fi}}{M_{cy}} \tag{8.17}$$

For overall buckling:

$$R = \frac{P_{fi}}{A_g p_c} + \frac{m M_{x,fi}}{M_b} + \frac{m M_{y,fi}}{p_y Z_y} \tag{8.18}$$

Equations (8.17) and (8.18) are based on the simple design method at ambient temperature (cf. equations (2.19) and (2.20)).

8.4.3.2   Eurocode 3 Part 1.2

EC3 also adopts the ambient temperature design method. The only difference is that the ambient temperature quantities are replaced by corresponding values at elevated temperatures.

8.5   FIRE RESISTANT DESIGN OF CONNECTIONS

The behaviour of connections in fire is complicated due to complex temperature distributions and interactions with the adjacent structure. Fortunately, since a connection is much "bulkier" than the connected members, its temperatures are much lower and it is rarely the weak link in the structure. Both Part 8 and EC3 treat the design of connections in fire in a very simple way. On connection fire resistance, Part 8 states:

> When fire protection materials are applied to a structure, the thickness of protection applied to a bolted or welded connection should be based on the thickness required for whichever of the members jointed by the connection has the highest section factor H_p/A.

EC3 states:

> The resistance of connections between members need not be checked provided that the thermal resistance $(d_f/\lambda_f)_c$ of the fire protection of the connection is not less than the minimum value of the thermal resistance $(d_f/\lambda_f)_m$ of the fire protection of any of the steel members joined by that connection, where d_f is the thickness of the fire protection material, λ_f is the effective thermal conductivity of the fire protection material.

Although the wording in Part 8 and EC3 is different, both statements are equivalent. However, since the section factor is much easier to calculate than the thermal resistance, the Part 8 statement is easier to implement in practice.

A potential problem arises in the specification of connection fire protection when one of the adjacent members, typically a beam, does not require fire protection. Under this circumstance, rapid heat conduction may occur via the connection from the unprotected beam to the protected column and this may damage the fire performance of the protected column. Future research is required in this area.

8.6   FIRE RESISTANT DESIGN OF COLD-FORMED THIN-WALLED STRUCTURES

The design methods described in Sections 8.3–8.5, apply to steel structures where local buckling does not present a problem. In thin-walled steel structures, local and distortional buckling is common and the behaviour of these structures in fire is complicated, making it difficult to develop simple design methods. The problem is made worse by the way these structures are heated in fire. Thin-walled steel structural members usually form part of a planar system (e.g. walling) where only one side of the structure is exposed to fire attack. This creates a steep temperature gradient through the thickness of the construction. Figure 3.5 in Chapter 3 shows an example of non-uniform temperatures in thin-walled panels exposed to fire attack on one side.

At present, since there is very little research study on this topic, design guidance on this type of steel structure is brief. EC3 simply states:

> For members with Class 3 cross-sections, other than tension members, it may be assumed that 4.2.1(1) (the load bearing requirement in fire) is satisfied if at time t the steel temperature T at all cross-sections is not more than 350 °C.

Part 8 does not place a limit on the type of cross-section. However, since design rules are based on those in BS 5950 Part 1 (BSI 1990a) which deals only with hot-rolled steel members where local buckling seldom occurs, the Part 8 design rules are not directly applicable to cold-formed thin-walled steel structures.

It appears that the fire performance of cold-formed thin-walled steel structures can only be evaluated by fire testing. This is indeed the case in a majority of applications. However, as cold-formed thin-walled steel members are becoming more widely used as main load-bearing structures in buildings, it is time for a suitable set of design rules to be developed for this type of structure. This section will introduce some recent developments in this area.

8.6.1   The Steel Construction Institute design guide

The SCI (Lawson 1993) has produced a design guide that is based on Part 8 for hot rolled steel sections, with some modifications of limiting temperatures to take into account the reduced strength of cold-formed steel at elevated temperatures.

The strength reduction factors at different strain levels for cold formed steel at elevated temperatures are given in Table 8.9. No results for steel stress at 2% strain are given since it is expected that cold-formed thin-walled members will fail at low levels of strain.

The limiting temperature method in Part 8 is adopted. The limiting temperatures for different types of thin-walled steel members are given in Table 8.10.

For beams supporting timber floors, it is assumed that timber floors will burn in fire and offer no protection to the steelwork. Therefore, the steel

Table 8.9 Strength retention factors for cold-formed steel at elevated temperatures

	Temperature (°C)								
	200	250	300	350	400	450	500	550	600
0.5% strain	0.95	0.89	0.83	0.76	0.68	0.58	0.47	0.37	0.27
1.5% strain	1.0	0.99	0.95	0.88	0.82	0.69	0.56	0.45	0.35

Source: Lawson (1993). Reproduced from *SCI Publication* P129, "Building design using cold-formed steel sections: fire protection", with the permission of The Steel Construction Institute, Ascot, Berkshire.

Table 8.10 Limiting temperatures of thin-walled cold-formed steel members

Member type	Limiting temperature at a load ratio of				
	0.7	0.6	0.5	0.4	0.3
Beams supporting concrete slabs	530	555	600	640	670
Beams supporting timber floors	450	485	530	575	625
Columns in walls	445	480	520	560	605
Slender columns	400	450	490	540	590
Other elements; studs and ties	400	450	490	540	590

Source: Lawson (1993). Reproduced from *SCI Publication* P129, "Building design using cold-formed steel sections: fire protection", with the permission of The Steel Construction Institute, Ascot, Berkshire.

structure is assumed to be at uniform temperature. Comparison between the steel strength at the 1.5% strain in Table 8.9 and the limiting temperature–load ratio relationship in Table 8.10 for this case reveals that they almost coincide, indicating a thin-walled steel beam is assumed to fail at 1.5% strain. This is shown in Figure 8.15.

When a beam supports concrete slabs, the temperature distribution in the steel beam will be non-uniform, giving a higher residual beam strength. In Section 8.3 on hot rolled steel beams, it has been pointed out that the bending moment capacity of a beam with uniform temperature distribution is about

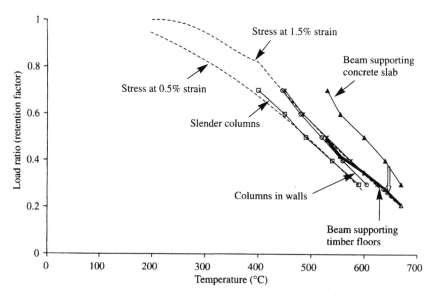

Figure 8.15 Limiting temperatures of cold-formed thin-walled structures (from Lawson 1993).

0.7 times that with non-uniform temperature distribution. Figure 8.15 suggests that this has also been applied to cold-formed thin-walled steel beams.

Columns are assumed to have uniform temperature distribution. Figure 8.15 compares the load ratio–limiting temperature relationships with the steel strength at 0.5% strain. It is clear that slender columns are assumed to fail at this level of strain. Columns in walls are assumed to have sufficient lateral restraint to reach the squash resistance. Therefore, they are expected to fail at higher strains and their limiting temperatures are slightly higher than those for slender columns.

The SCI method cannot be used to deal with columns with non-uniform temperature distributions. A limited study by Wang and Davies (2000) indicates that the ambient temperature design method in Eurocode 3 Part 1.3 (CEN 1996) may be used. However, the column should be designed as a beam-column subjected to combined axial load and bending moments. The bending moments are a result of the axial load acting upon the column thermal bowing and shift of the centroid. The shift of the centroid is due to non-uniform distribution of stiffness in the cross-section and the thermal bowing deflection may be calculated using equation (5.21). The effective width in fire may be taken as the same as that at ambient temperature (cf. Section 5.2.1).

8.6.2 Temperatures in thin-walled steel structures

An important difference between the behaviour of a hot-rolled and a cold-formed thin-walled steel structure is the difference in their temperature distributions. For a hot-rolled structure, the structural member is fully exposed to fire attack and it is also massive. Therefore, uniform temperature distribution within an element may be assumed and equations (6.40) and (6.45) can be used.

A cold-formed thin-walled steel member is usually protected by planar or flat systems as shown in Figure 8.16 and fire exposure is on only one side of

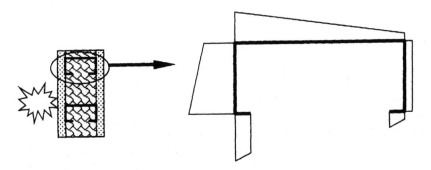

Figure 8.16 Temperature distribution in a thin-walled section exposed to fire on one side.

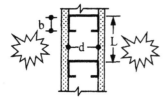

Figure 8.17 Section factor for a cold-formed panel exposed to fire on two sides.

the system. Therefore, a temperature gradient is developed in the thickness direction of the system. In addition, because the steel section is thin, heat losses from the steel surfaces are high so that there is also a temperature gradient in the width direction of the thin-walled steel section. The lumped mass method in equations (6.40) and (6.45) are no-longer applicable and numerical heat transfer analysis becomes necessary.

The lumped mass method is only applicable in a very limited way. If a thin-walled steel section is protected by a planar system and is heated from both sides, as shown in Figure 8.17, the section factor may be calculated using the following equation:

$$\frac{A_p}{V} = \frac{L}{(2b+d)t}r \tag{8.19}$$

where the first term on the right hand side of equation (8.19) is the conventional section factor and r is a modification factor to take into account the effect of heat losses from steel surfaces to the air void. Modification factor r is given by

$$r = 1 - 0.0003\frac{bd}{A_s} \tag{8.20}$$

8.7 STAINLESS STEEL STRUCTURES

Due to architectural demand and superior corrosion resistance, stainless steels are becoming more widely used. Although their fire resistance is only a minor factor in determining whether to use stainless steel or not, stainless steel does have superior fire resistance to conventional carbon steels.

Figure 8.18 compares strength retention factors of conventional and stainless steels at different elevated temperatures (Baddoo 1999). Whilst conventional carbon steel loses about 50% of its strength at a temperature of around 600 °C, the temperature which gives the same loss in the strength of stainless steel is much higher, at about 800 °C. Also the surface of stainless

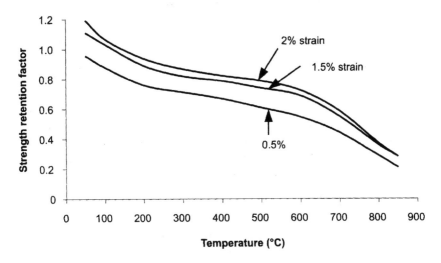

Figure 8.18 Retention factors of stainless steel (*The Structural Engineer*. "Stainless steel in fire", **77**(19), Badoo, N. (1999). Reproduced with the permission of the Institution of Structural Engineers).

steel has a much higher reflectivity, hence the emmissivity of stainless steel is much lower. Typically, the emmissivity of stainless steel is about 0.3–0.4, compared to about 0.8 for conventional steel (Thomas 1980). This gives a much lower temperature in a stainless steel structure in fire. For realistic loading conditions, it is almost certainly that stainless steel structures will be able to be engineered to provide sufficient fire resistance without the need of fire protection.

With suitable modifications to take into consideration the reduced strength and stiffness of stainless steel at elevated temperatures, the same design method for conventional steel may be extended to stainless steel structures (Baddoo and Burgen 1998).

8.8 OTHER TYPES OF STEEL STRUCTURES

The design methods described in Sections 8.3–8.5 are for general use on steel members in multi-storey steel framed buildings where flexural bending is the dominant mode of structural behaviour. A number of separate design methods have also been developed for fire safety of steel structures of more specialist natures. They are portal frames, water cooled structures and external steelwork. Consultation should be made to the specialist design documents that have been produced for these types of steel structures. This section will only provide a brief description of these specialist problems and list the main assumptions adopted in these design documents.

8.8.1 Portal frames (Newman 1990)

Portal frames are usually single storey buildings with a small density of occupants. Escape in the case of fire attack is relatively easy. Therefore, fire safety of a portal frame only becomes a requirement when the portal frame is adjacent to another building and it is necessary to prevent fire spread from the portal frame building to the adjacent building. Fire spread from a portal frame building is usually through collapsed walls, consequently, the structural safety requirement for a portal frame under fire attack is to ensure that the portal frame columns in the walls remain upright so that the walls are stable. The portal frame girders may be allowed to collapse. Other fundamental design assumptions are:

- At collapse, plastic hinges develop at the eaves locations (column/girder junctions);
- Due to large deformation, the portal frame girder is in a state of catenary action that produces an overturning moment at the column base;
- Only a small part of the roof loading needs to be included in the fire limit state;
- Provided the nominally simple column bases have sufficient moment capacities to resist the overturning bending moments, the portal frame columns may be assumed to remain stable.

Using the Newman (1990), portal frames are usually designed without fire protection.

8.8.2 Water cooled structures (Bond 1975)

The design of water cooled structures relies on the principle that boiling water has a very high value of convective heat transfer coefficient, some three or four orders higher than the convective heat transfer coefficient of air. If boiling water is replenished, heat on a water cooled steel structure is taken away and steel temperatures remain low. Typically, steel temperatures in a water cooled structure do not exceed 150 °C. Therefore, structural stability of a water cooled structure is rarely a design issue. Design is mainly concerned with hydraulic calculations to ensure sufficient water supply and circulation in case of a fire.

Water cooling a steel structure to achieve fire protection is expensive and because of this, it is rarely used solely for the fire protection purpose. It is usually combined with other functions. An excellent example (Bressington 1997) of recent application of this technique is in the roof truss of the cargo handling facility of Hong Kong Air Cargo Terminals Ltd. The steel structural roof truss is made of circular hollow sections and is used as the water distribution pipe for sprinklers. On operation of the sprinklers in the event of a fire, internal water flow through the steelwork members also provides sufficient cooling.

8.8.3   External steelwork (Law and O'Brien 1989)

In the case of a building enclosure fire, fire exposure on the external steelwork differs from that on the interior steelwork in two ways:

1   The fire temperature to the external steelwork is much lower than that of the interior steelwork;
2   The external steelwork may not be directly engulfed in fire.

These two differences ensure that temperatures in the external steelwork are kept lower than their failure temperatures so that fire protection is not necessary.

8.9   COST EFFECTIVE DESIGN OF STEEL STRUCTURES FOR FIRE PROTECTION

In the design of a steel structure, design checks for the ambient temperature condition and for fire resistance are usually separated. From the ambient temperature design, the lightest steel section size is selected. If the full load is applied at ambient temperature, the load ratio for fire design is about 0.6, giving the maximum limiting temperature of about 700 °C according to Part 8 (cf. Tables 8.1 and 8.7). On the other hand, Figure 8.1 indicates that even for a standard fire resistance rating of 30 min, the temperatures attained in unprotected steelwork are usually higher than 700 °C. Therefore, if the design method in Part 8 or EC3 is followed, fire protection will almost certainly be necessary. When fire protection is unavoidable, it is sometimes useful to consider using a heavier steel section so as to minimize the total cost of structural steel and fire protection.

An example is given below to illustrate the benefit of combining ambient and fire design calculations for a steel beam. In this example, fire protections are either by sprayed protection or intumescent coating. Design data for the steel beam are:

Span:	$L = 6$ m
Loading:	Dead load $= 30$ kN/m, Live load $= 15$ kN/m
Steel grade:	S275
Lateral and torsional restraint:	Full restraint provided by floor slabs
Fire resistance:	60 min of standard fire exposure

Table 8.11 Example thickness of required intumescent coating

$A_p/V(m^{-1})$	Dry film thickness (μm) at a limiting temperature of				
	620 °C	670 °C	720 °C	770 °C	820 °C
0–170	675	600	525	450	400
171–190	775	700	625	550	505
191–250	875	800	725	650	600

Source: Yandzio et al. (1996). Reproduced from SCI Publication 160, "Structural fire design: off-site applied thin film intumescent coatings", with the permission of The Steel Construction Institute, Ascot, Berkshire.

For an ambient temperature design (according to BS 5950 Part 1):

$$w = 1.4 \times 30 + 1.6 \times 15 = 66 \, \text{kN/m} \qquad M = 1/8 \times 66 \times 6^2 = 297 \, \text{kN.m},$$

Required $S_x = 297 \, (\text{kN.m})/275 \, (\text{N/mm}^2) = 1080 \, \text{cm}^3$, select $457 \times 152 \times 52$ UB, giving $S_x = 1096 \, \text{cm}^3$ and $M_p = 0.275 \times 1096 = 301.4 \, \text{kN.m}$

For fire design (according to BS 5950 Part 8):

$$w = 1.0 \times 30 + 0.8 \times 15 = 42 \, \text{kN/m} \qquad M = 1/8 \times 42 \times 6^2 = 189 \, \text{kN.m}$$

The load ratio is $189/301.4 = 0.627$, giving a limiting temperature of 612°C (Case 3 of Table 8.1)

Using spray protection, the cost of fire protection is about £10.89/m² (DLE 1997). Total fire protection cost is £10.89 × L × H_p, where $L = 6$ m and H_p (perimeter length of the cross-section) = 1.485 m, giving the cost of fire protection at £97.03. Assume that the fabricated steel structural cost is £900/ton, the cost of steel is £282.42. The total cost of structural steel and fire protection is £369.45.

The thickness of intumescent coating is provided by intumescent coating manufacturers. Table 8.11 shows a typical example. The section factor for this beam is 200 m⁻¹. Extrapolating the information in Table 8.11, the required dry film thickness is about 887 µm.

The cost of intumescent coating is also provided by the manufacturer. Figure 8.19 shows a typical example. For a coating thickness of 887 µm, the cost is £24.97/m². The cost of intumescent coating is £199.60, giving the total cost of structural steel and fire protection at £482.02.

Using the same procedure, the total cost of the steel beam and fire protection is calculated for a few other UB sections that have higher bending resistance. The results of these calculations are given in Table 8.12.

In Table 8.12, the same unit cost is used for the spray fire protection regardless of the limiting temperature of the steel beam. Since a significant

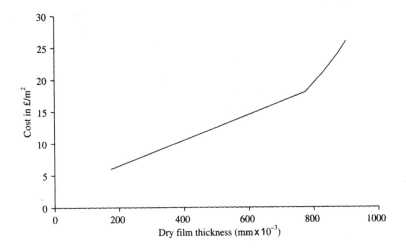

Figure 8.19 Illustrative price of intumescent coating (from Yandzio *et al.* 1996). Reproduced with the permission of The Steel Construction Institute, Ascot, Berkshire.

Table 8.12 Comparisons of the total cost of steel structure and fire protection

UB section	Steel cost (£)	Spray protection		Intumescent coating protection		
		Cost (£)	Total cost (£)	Thickness (μm)	Cost (£)	Total cost (£)
457 × 191 × 67	362.34	97.97	457.31	615	129.11	491.45
457 × 152 × 67	362.88	89.49	452.37	618	122.09	484.97
457 × 152 × 60	322.92	87.13	410.05	745	139.27	**462.19**
457 × 152 × 52	282.42	87.03	**369.45**	887	199.60	482.02
406 × 178 × 67	362.34	86.59	448.93	635	120.83	483.17

part of the spray fire protection cost is labour cost, a small change in the fire protection thickness has negligible influence on the fire protection cost. The total cost gives the combined structural and fire protection cost. From Table 8.12, it can be seen that although the fire protection cost of the lightest section (457 × 152 × 52 UB) is marginally higher than that of another section (406 × 178 × 67 UB), it still gives the lowest total cost.

For intumescent coating, the lightest steel section requires the highest dry film thickness. Since it also has a very high surface area to be protected, its fire protection cost is much higher than other sections. As a result, the lightest steel section does not give the lowest total cost.

The absolute fire protection cost values in Table 8.12 cannot be used in other cases. However, a general trend can be observed. If spray or board protections using conventional passive fire protection materials are used, the lightest steel section usually gives the lowest total cost. If intumescent coating is used for fire protection, it is useful to consider using heavier sections to reduce the total cost.

Of course, if fire protection is required, the total cost of the steel construction will be high. In the example shown above, the spray fire protection adds about 30% to the cost of the steel structure and the intumescent coating about 40%. In order to increase the competitiveness of a steel structure, it is best to completely eliminate fire protection. This is the topic of Chapter 10.

8.10   FEASIBILITY OF USING CATENARY ACTION TO ELIMINATE FIRE PROTECTION IN BEAMS

In Section 5.5.3, it was suggested that catenary action could occur in steel beams under fire. Obviously, further research studies are required to investigate the reliability of this load carrying mechanism, in particular, the magnitude of catenary force that a steel beam may impose on the adjacent structure. The following example investigates possible savings in the fire protection cost if catenary action were reliable and the theory in Section 5.5.2 were correct.

Consider a plane steel frame of five bays and four storeys. Each storey is 4 m high. The beam span and loading are as in the previous example. Assume simple construction. Consider the bottom floor structure only. If fire resistant design proceeds according to Part 8 as in the previous example, results in Table 8.13 may be obtained.

All steel sections will need fire protection and the total weight of protected steel is 3064 kg. The total cost of steel and fire protection is £3860.4.

If catenary action is used, the two exterior columns will have to be increased to resist the catenary forces in the unprotected steel beams. The internal columns do not need change because the floor slabs will be able to transfer the catenary forces to the edge columns.

All beams will be unprotected. From Figure 8.1, the unprotected steel temperature at 60 min of the standard fire exposure is 937 °C, giving a strength retention factor of 0.0546 from Part 8. Assume that the unprotected steel beam is in pure catenary action, giving a catenary force of $0.0546 \times 0.275 \times 6660 = 100 \, \text{kN}$. From equation (5.22), the maximum

Table 8.13 Cost of an artificial steel frame, without considering catenary action

Element	Size	Steel cost (£) (@£900/ton)	Fire protection cost (£@£10.89/m²)	Total cost (£)
Beam (5 off) from previous example	457 × 152 × 52 UB 282.4		97	369.4
Internal columns (4 off)	203 × 203 × 71 UC 255.6		53.1	308.7
External columns (2 off)	203 × 203 × 46 UC 165.6		52.4	218

deflection in the beam may be found to be $(42 \times 6^2)/(8 \times 100) = 1.89$ m. This deflection is far too high, but we proceed with the other calculations.

For information, the interior columns may be designed as before. The edge columns will have to be increased to accommodate a lateral load of 100 kN and the original axial load. The lateral load produces a bending moment of 100×8 m/4 = 200 kN.m in the column. If the new edge columns are designed to have a comparable load ratio (≈ 0.5) as before, the required edge column size is $254 \times 254 \times 89$ UC. The associated steel and fire protection costs are £320.4 and £66.5, respectively.

The total cost of steel and fire protection is now £3732, producing a saving of about 3.4%. Only columns will need fire protection and all steel beams will be unprotected. Therefore, the total protected steel weight is 1848 kg and the unprotected steel weight is 1560 kg. Obviously, the saving will be higher if the beams are longer and on a large scale project, the saving can become quite impressive.

8.11 CONCLUDING REMARKS

This chapter has introduced the design of steel structures for fire resistance. It has covered the design of hot-rolled steel beams, columns, connections and frames and a limited number of cold-formed thin-walled steel members. Both Part 8 and EC3 deal with this subject in a similar way to that of their parent codes at ambient temperature. Results of a number of comparative studies in this chapter indicate that both design methods have similar accuracy, with EC3 tending to involve more calculations. A number of modifications have been suggested to improve the accuracy of both methods. An example is given to show how the total cost of structural steel and fire protection may be minimized by considering the ambient temperature and fire designs together. Current design rules are based on the flexural bending behaviour of a structural member at ambient temperature. Another example is given to illustrate how the total cost of fire protection may be further reduced by considering catenary action in unprotected steel beams in fire. Obviously, further research studies are required to investigate the reliability of this load carrying mechanism.

Chapter 9

Design of composite structures for fire safety

The behaviour of steel and composite structures in fire are similar, therefore design methods for steel and composite structures in fire have many similar concepts and features. However, there is one significant difference in their behaviour; due to complicated temperature distributions in a composite structure, it is time consuming to implement in practice some of the general structural engineering principles given in design codes. Therefore, this chapter has the following two objectives:

1 to present the general FR design methods for principal composite structural members; and
2 to present, wherever possible, simplified calculation procedures for the implementation of the general design methods.

This chapter will present design methods for composite slabs, composite beams and composite columns. Since BS 5950 Part 8 (BSI 1990b, hereafter to be referred to as Part 8) has a very limited coverage on composite members, discussions will mainly be based on design methods in Eurocode 4 Part 1.2 (CEN 2001, hereafter to be referred to as EC4).

9.1 COMPOSITE SLABS

Composite slabs are constructed from reinforced concrete slabs in composite action with steel decking underneath. The steel decking acts as support to the concrete during construction and is generally profiled to maximize structural efficiency. Composite slabs usually form the floor of a FR compartment. Hence, they should meet all the requirements of FR construction, i.e. in addition to sufficient load bearing resistance, they should also have adequate insulation and maintain their integrity during fire attack.

Composite floor slabs are non-combustible and will not suffer integrity failure within the slabs. However, the problem of integrity failure may occur at the junctions between a composite slab and other construction elements.

Table 9.1 Minimum slab thickness to satisfy insulation condition

Standard fire resistance time (minutes)	30	60	90	120	180	240
Minimum slab thickness (mm)	60	80	100	120	150	174

Source: CEN (2001). Reproduced with the permission of the British Standards Institution under licence number 2001SK/0298.

Particular attention should be paid to the slab edges where large cracks may occur due to large rotations. It is important that reinforcement bars should be made continuous over the supports.

To check whether a slab can fulfill the insulation requirement, it is necessary to carry out a heat transfer analysis to determine temperatures on the unexposed surface of the slab. Results of this temperature analysis can then be used to determine the minimum slab thickness above which the unexposed surface temperature is unlikely to cause further ignition. The minimum slab thickness for each realistic fire condition will be different. For design under the standard fire exposure, Table 9.1 gives the recommended minimum slab thickness in EC4.

For most applications where the required standard fire resistance rating does not exceed 90 min, the required minimum slab thickness will almost certainly be less than that required by other functions such as control of deflection. Hence, the insulation requirement for composite slabs is very rarely a problem in FR design.

9.1.1   Load bearing capacity of one-way spanning composite slabs

When calculating the load bearing resistance of a composite slab in fire, it is often assumed that it is one-way spanning, being effective only in the direction of the concrete rib. For a continuous composite slab, the plastic design method may be used. In this method, plastic hinges are assumed to form at the supports and locations of the maximum bending moment in the span of the slab. Design conditions to be satisfied according to Section 2.7 are given below.

9.1.1.1   *For the interior span of a continuous slab under uniformly distributed load*

$$M_{+,\text{fi}} + M_{-,\text{fi}} \geq M_{\text{fi,max}} \quad \text{i.e.} \quad M_{+,\text{fi}} + M_{-,\text{fi}} \geq \frac{1}{8}wL^2 \tag{9.1}$$

where $M_{+,\text{fi}}$ and $M_{-,\text{fi}}$ are the sagging and hogging bending moment resistance of the slab respectively and $M_{\text{fi,max}}$ is the maximum free bending moment in the slab under fire conditions.

9.1.1.2 For the end span of a continuous slab with uniformly distributed load

$$M_{+,\mathrm{fi}} + 0.45M_{-,\mathrm{fi}} \geq \frac{1}{8}wL^2 \tag{9.2}$$

In order to determine the slab load carrying capacity in fire, the sagging bending moment capacity $M_{+,\mathrm{fi}}$ and hogging bending moment capacity $M_{-,\mathrm{fi}}$ should be evaluated.

9.1.2 Sagging bending moment capacity $M_{+,\mathrm{fi}}$

When calculating the sagging bending moment capacity of a slab, the reinforcement near the fire side is in tension and the concrete in compression is on the unexposed side of the slab. Since the temperature rise on the unexposed side of the slab is required to be below 140 °C to fulfill the insulation requirement, the concrete in compression can be assumed to be cold and its cold strength may be used when calculating the sagging bending moment capacity of the slab. Contributions from the steel decking are usually ignored because the decking will be unprotected and will debond when under direct fire attack. Figure 9.1 shows the calculation procedure. In Figure 9.1, f_c is the design strength of concrete in bending at ambient temperature; A_r, $p_{y,r}$ and $k_{y,r}(T)$ are the area, design strength at ambient temperature and strength reduction factor of the reinforcement at temperature T and $k_{y,r}(T)$ may be obtained from Figure 9.2.

For design under the standard fire exposure, Figure 9.3 may be used to obtain the reinforcement temperature. If design is for realistic fire exposures, numerical thermal analyses will be necessary.

9.1.3 Hogging bending moment capacity $M_{-,\mathrm{fi}}$

Under a hogging bending moment, the compression face of a composite slab is exposed to fire where there is a very steep temperature gradient. When

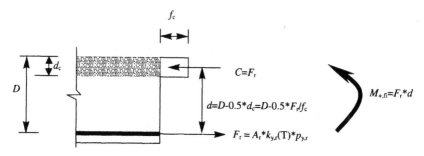

Figure 9.1 Calculation method for sagging bending moment capacity.

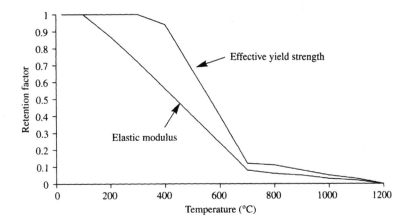

Figure 9.2 Retention factors of cold worked reinforcing steel (from CEN 2001). Reproduced with the permission of the British Standards Institution under licence number 2001SK/0298.

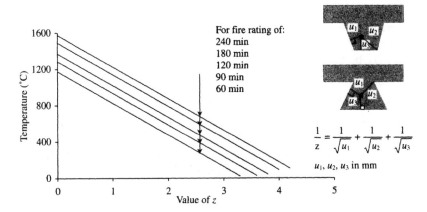

Figure 9.3 Temperatures of reinforcing bars (from CEN 2001). Reproduced with the permission of the British Standards Institution under licence number 2001SK/0298.

calculating the hogging bending moment capacity, the composite slab, including the concrete in the ribs, should be divided into a number of layers each of approximately constant temperature. The contribution of each layer should be evaluated separately and then integrated to give the total slab resistance.

Clearly, this calculating procedure is tedious. To simplify calculations, a ribbed slab may be converted into a flat slab with an effective flat slab depth. For a common trapezoidal decking, the effective flat slab depth is equal to the net slab thickness plus half the rib depth. For design under the

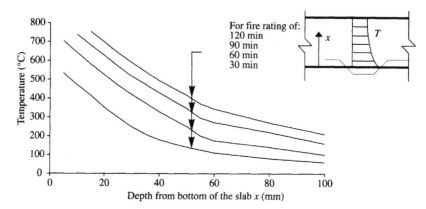

Figure 9.4 Temperatures of a concrete slab exposed to the standard fire from underneath (from CEN 2001). Reproduced with the permission of the British
Standards Institution under licence number 2001SK/0298.

standard fire exposure, slab temperatures at different depths may be
obtained from Figure 9.4. Values in Figure 9.4 apply to normal weight
concrete. Lightweight concrete has lower temperatures due to a lower thermal
conductivity. As an approximation, temperatures in a lightweight concrete
slab may be taken as 0.9 times those in Figure 9.4.

9.1.4   Load bearing capacity of two-way spanning slabs

The load carrying capacity of a composite slab in one-way spanning is usually
sufficient under fire conditions. However, in some cases, it may be necessary
to utilize the slab strength in two-way spanning. This is particularly the case
when it is necessary to justify the elimination of fire protection from some of
the slab-supporting steel beams. Section 5.6.1 has briefly introduced the yield
line analysis for reinforced concrete slabs and this method may be used to give
a safe estimate of the slab load carrying capacity.

The real benefit of utilizing the strength of a slab in two-way spanning is
the possibility of using tensile membrane action in the slab, under which the
strength of the slab can be many times higher than that given by yield line
analysis. Section 5.6.4 has described this load carrying mechanism in some
detail. For detailed design equations and examples, reference should be
made to the papers by Bailey and Moore (2000a,b).

Section 5.6.4 has pointed out a number of important assumptions that
have been adopted in developing the design equations based on tensile
membrane action. It is important to observe the implications of these
assumptions and to ensure that these assumptions are fulfilled in the real
structure under consideration.

9.2   COMPOSITE BEAMS

Composite beams usually form part of a floor slab, with the floor slab providing lateral supports to the top flange of the steel section. Also a composite beam is usually designed as simply supported. Since the compression flange of the steel section is restrained, LTB of the steel section does not occur. Therefore, the FR design of a composite beam is essentially confined to ensure that the maximum applied bending moment in the beam does not exceed the plastic bending moment capacity of the composite cross-section under fire conditions.

9.2.1   BS 5950 Part 8

The limiting temperature method can also be used for composite beams and Table 9.2 gives the limiting temperatures of composite beams with complete shear connection and partial shear connection. The load ratio should be taken as $M_{max,fi}/M_p$, where $M_{max,fi}$ is the maximum bending moment in the beam in fire and M_p the plastic bending moment capacity of the composite beam, calculated using equations (2.23) and (2.25).

Table 9.2 Limiting temperatures of composite beams

Conditions of application	Limiting temperature (°C) at a load ratio of						
	0.7	0.6	0.5	0.4	0.3	0.2	0.1
Members in bending supporting and acting compositely with concrete slabs or composite slabs with filled voids:							
1 unprotected members, or protected members with fire protection that can undergo large strains							
– 100% degree of shear connection	550	580	610	645	685	740	840
– 40% degree of shear connection	575	600	635	665	700	760	860
2 other protected members							
– 100% degree of shear connection	505	540	575	620	660	715	790
– 40% degree of shear connection	535	565	605	640	675	736	800

Source: BSI (1990b). Reproduced with the permission of the British Standards Institution under licence number 2001SK/0298.

9.2.2   Eurocode 4 Part 1.2

9.2.2.1   *Conventional composite beams*

For a conventional composite beam with concrete slabs on top of the steel section, the sagging bending moment capacity of the composite cross-section may be calculated by the bending moment capacity method as illustrated in

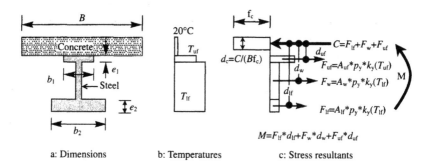

a: Dimensions   b: Temperatures   c: Stress resultants

Figure 9.5 Method of calculating the plastic bending moment capacity of a composite cross-section in fire.

Figure 9.5. The concrete in compression may be assumed to be cold and the steel temperatures may be calculated using equations (6.40) or (6.45). If the steel section is protected, it may be assumed to have a uniform temperature distribution and its section factor A_p/V is that of the entire section, calculated according to Table 6.5. If the steel section is unprotected, the steel section will have a non-uniform temperature distribution. For temperature calculations, the steel section may approximately be divided into two parts: (1) the upper flange; and (2) the lower flange plus the web. In Figure 9.5, the section factors of these different parts can be calculated using the equations:
For the lower flange and web:

$$\frac{A_p}{V} = \frac{2(b_2 + e_2)}{b_2 e_2} \qquad (9.3)$$

For the upper flange:

$$\frac{A_p}{V} = \frac{b_1 + 2e_1}{b_1 e_1} \qquad (9.4)$$

Once the steel temperatures are calculated, the residual tensile strength of each part of the steel section is calculated using the strength retention factors at the appropriate temperature. If the PNA is in the concrete flange, Figure 9.5 illustrates the calculation procedure for the plastic bending moment capacity of the composite cross-section.

As a further simplification, the steel section may be assumed to be at the same temperature as that of the lower flange. The plastic bending moment capacity of the composite cross-section with non-uniform temperature distribution is obtained by increasing that with uniform temperature by multiplying by a factor of 1/0.9.

This amplification factor of 1/0.9 for a composite beam is directly comparable to a factor of $\kappa_1 = 0.7$ in equation (8.2) for a steel beam with non-

Table 9.3 A comparison of plastic bending moment capacities obtained using different calculation methods for the composite cross-section in Figure 9.6

Lower flange temperature (°C)	600 °C			800 °C		
Upper flange temperature (°C)	600	450 ($\Delta T = 150$)	350 ($\Delta T = 250$)	800	650 ($\Delta T = 150$)	550 ($\Delta T = 250$)
From the bending moment capacity method (kN.m)	577.36	611.45	621.3	138.22	166.44	197.98
From using $\kappa = 0.9$ (kN.m)	N/A	641.51	641.51	N/A	153.58	153.58
From the limiting temperature method (kN.m)	N/A	634	634	N/A	166.46	166.46

uniform temperature distribution. For a composite beam, since the upper flange is close to the plastic neutral axis of the composite cross-section, a lower temperature in the upper flange does not have as a pronounced effect on the plastic bending moment capacity of the composite cross-section as on a purely steel cross-section.

If the composite beam does not have complete shear connection, the amplification factor may be replaced (Lawson and Newman 1996) by

$$\kappa = 0.7 + 0.2\,K \tag{9.5}$$

where K is the degree of shear connection at ambient temperature.

Results in Table 9.3 have been prepared to check the accuracy of the simplified method. Input data for these calculations are shown in Figure 9.6. Table 9.3 also includes results calculated using the Part 8 limiting temperature method.

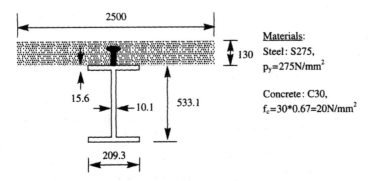

Figure 9.6 Dimensions of a composite beam used in calculations for Table 9.3.

Two observations follow from the results in Table 9.3. Firstly, the lower flange temperature can affect whether non-uniform temperature distribution has any effect on the plastic bending moment capacity of the composite cross-section. The difference between different non-uniform temperature distributions is small if the lower flange temperature is low (600 °C), but can be quite high (15%) if the lower flange temperature is high (800 °C). Secondly, using the limiting temperature method of Part 8 produces similar results to using the amplification factor of 1/0.9. Both seem to give a lower (safe) bound solution.

Equation (9.5) is obtained on the assumption that the upper flange is in contact with a solid concrete slab or a composite slab with filled voids so that the concrete/steel interface is protected from fire attack. However, if the composite slab is profiled and the voids are not filled, the section factor of the upper flange should be calculated using

$$\frac{A_p}{V} = \frac{2(b_1 + e_1)}{b_1 e_1} \tag{9.6}$$

In this case, it is very likely that the temperature distribution in the steel section will be near to uniform. As a simplification, calculation of the plastic bending moment capacity of a composite cross-section with unfilled voids may be based on uniform temperature distribution at the lower flange temperature.

9.2.2.2   Partial fire protection to the steel section in a composite beam

In conventional fire protection, the entire steel cross-section of a composite beam is protected. However, it has to be appreciated that for a composite beam, the most important part of the steel cross-section is the lower flange. Therefore, fire protection of the lower flange only can be an effective means of reducing the cost of fire protection. Obviously, this method is best suited to spray fire protection where the profile of fire protection can be easily changed.

An interesting study has been carried out by Wang (1998a) to investigate the feasibility of partial fire protection in a composite beam. It was found that by protecting only the lower flange and a small portion of the web (1/4) of the steel section, temperatures in the lower flange were such that the composite beam still had sufficient load carrying capacity to resist the applied load in fire. A method has been proposed to calculate temperatures in the lower flange of the beam with partial fire protection. The additional heat conducted from the unprotected web to the lower flange gives an increase in the temperature of the lower flange in a partially protected beam compared to a completely protected beam, which necessitates a thick fire protection.

However, because of the reduced fire protection area, the total weight of fire protection material in the partially protected beam is often still much less than that in a completely protected beam. However, the main advantage is that only a small area of the steel section needs fire protection (typically 1/3 of that in complete protection), saving substantially on labour cost.

9.2.2.3 Partial shear connection

Partial shear connection is best dealt with using the limiting temperature method in Part 8. Otherwise, information will be required of temperature rises in shear connectors and degradations of the strength of shear connectors at elevated temperatures, which are generally difficult to obtain. The limiting temperatures for partial shear connection are given in Table 9.2. They are based on observations of standard fire tests on composite beams with partial shear connection (Newman and Lawson 1991). It can be seen from Table 9.2 that the limiting temperatures of a composite beam with partial shear connection are higher than those with complete shear connection. This is because the degree of shear connection is based on the definition at ambient temperature (equation (2.24)). Under fire conditions, shear connectors have lower temperatures than the steel section, thereby effectively increasing the degree of shear connection in fire.

9.3 COMPOSITE COLUMNS

9.3.1 Resistance to axial load according to EC4

The calculation method in EC4 for composite columns is very similar to that in Eurocode 3 Part 1.2 for steel columns described in Section 8.4.1. Obviously, the squash load and stiffness of the composite cross-section should be used. The general equation for calculating the squash load of a composite column is

$$P_{u,fi} = \sum (A_i f_i)_c + \sum (A_i f_i)_s + \sum (A_i f_i)_r \tag{9.7}$$

where f_i is the design strength of the ith layer and subscripts "c", "s" and "r" represent concrete, steel and reinforcement respectively. Due to non-uniform temperature distribution, each component of the composite cross-section is divided into a number of layers of approximately the same temperature.

Similarly, the rigidity (EI) of the composite cross-section is calculated using the following equation

$$(EI)_{fi} = \sum \varphi_s (EI)_{i,s} + \sum \varphi_c (EI)_{i,c} + \sum \varphi_r (EI)_{i,r} \tag{9.8}$$

where symbols E and I are the initial modulus of elasticity and second moment area of the appropriate component material.

The empirical factor φ allows for the material non-linear stress–strain relationships at elevated temperatures, its value being dependent on the final stress of the material. In EC4, the value of φ for concrete is 0.8. For steel and reinforcement, this value is close to 1.0, depending on the standard fire resistance time.

Equation (9.8) is clearly difficult to implement in practice if calculations have to consider different values of φ for different parts of the composite cross-section under different fire exposure conditions. It is proposed that design calculations should use constant values of $\varphi = 0.8$ for concrete and 1.0 for steel and reinforcement. The proposed simplification is reasonable because:

1 the EC4 factors are empirical and are obtained for the standard fire condition only and cannot be more widely applied to realistic fire exposures;
2 inaccuracy in mechanical properties of steel and concrete at elevated temperatures overwhelms that caused by using inaccurate values for the φ factors; and
3 the proposed simplification is consistent with the ambient temperature approach.

9.3.2 Simplified calculation methods for concrete filled columns

Once values of the squash load $P_{u,fi}$ and rigidity $(EI)_{fi}$ of a composite cross-section are obtained, it is a simple task to use the calculation method in Section 8.4.1 to obtain the strength of the composite column. However, it is a tedious process to obtain these two values and their calculations involve the following two lengthy steps:

1 the non-uniform temperature distributions in the composite cross-section have to be evaluated using numerical methods; and
2 due to the non-uniform temperature distribution, the composite cross-section has to be divided into a number of fine layers of approximately the same temperature so as to calculate their contributions to the squash load and rigidity of the composite cross-section.

In this section, some simplified methods are described for concrete filled columns subject to the standard fire exposure. These approximate methods are given both for temperature calculations and also for directly obtaining the squash load and rigidity of a composite cross-section without calculating temperatures. The latter involves less calculation, but the former may give more accurate results.

9.3.2.1 Temperature calculation method

The temperature calculation method is based on the method of Lawson and Newman (1996) with minor modification by Wang (2000a). This method assumes that the composite column is unprotected.

In this method, the steel shell temperature is calculated by

$$T_s = C_2 T_{fi} \tag{9.9}$$

where T_{fi} is the standard fire temperature and C_2 is a multiplication factor depending on the fire resistance time. C_2 is given by

$$C_2 = 1 - 0.02t^1 \frac{120 - FR}{120} \quad \text{but} \quad C_2 \leq 1.0 \tag{9.10}$$

where FR is the fire resistance rating in minutes and t the thickness of the steel shell in mm. The concrete temperature is calculated by

$$T_c = C_1 C_2 T_{slab} \tag{9.11}$$

where C_1 is a multiplication factor depending on the composite section size and locations of the concrete (and is independent of the standard fire resistance time) and T_{slab} is the temperature in an infinitely wide concrete slab exposed to fire on one side (given in Figure 9.4). Values of C_1 are given in Table 9.4.

If reinforcement is used, the reinforcement temperature should be taken as that of the concrete at the same location. Furthermore, for reinforcement in a square section, due to heating from two sides of the composite section, the reinforcement temperature should be calculated from an equivalent depth of half the concrete cover depth.

9.3.2.2 Direct strength and stiffness calculation methods

In deriving a simplified method to directly obtain the squash load and rigidity of a concrete filled column, it is assumed that the composite column

Table 9.4 Multiplication factor C_1

Diameter or size of square section in mm	Distance of centre of layer from out surface in mm				
	10	30	50	70	>70
200	1.08	1.22	1.41	1.60	1.80
300	1.05	1.14	1.22	1.36	1.50
400	1.03	1.09	1.18	1.25	1.35
500	1.02	1.07	1.12	1.18	1.25

Source: Lawson and Newman (1996). Reproduced from SCI Publication 159, "Structural fire design to EC3 & EC4, and comparison with BS 5950", with the permission of The Steel Construction Institute, Ascot, Berkshire.

is either unprotected or protected but unreinforced. This is a reasonable assumption because unprotected unreinforced concrete filled steel tubular columns can already achieve a high level of fire resistance. To improve its fire resistance further, either reinforcement or fire protection may be used, with offsite intumescent coating being the preferred choice as it retains the architectural appearance and takes fire protection out of the critical construction path. External protection and reinforcement may be applied together, but it is seen that this combination would result in a complicated form of structure that loses much of the appeal of concrete filled steel tubular columns.

The simplified method has been developed based on results of comprehensive numerical studies (Wang 1997c, 2000a; Wang and Kodur 1999) using finite element methods for temperature calculations and equations (9.7) and (9.8) for the strength and rigidity of a composite cross-section, and wherever possible, validated by available experimental results.

Unprotected columns

For an unprotected column, design parameters are the steel tube size, the standard fire resistance rating; and the grades of steel and concrete. It is assumed that the number of available steel tube sizes is finite and that the standard fire resistance rating is expressed in only a few multiples of 30 min. Taken collectively, the only design variable is the combination of grades of steel and concrete. If information is available for the squash load and rigidity of the entire set of structural steel tubes under different standard fire resistance ratings for one combination of grades of steel and concrete, the simplified method relates the squash load and rigidity of a composite column with a different combination of grades of steel and concrete to those with the reference combination grades of steel and concrete.

The relationship for the squash load of a composite cross-section is

$$\frac{P_{u,fi}}{P_{u,fi,m=0}} = \frac{f_c}{f_{c,m=0}} + \left(\frac{P_{u,a}}{P_{u,a,m=0}} - \frac{f_c}{f_{c,m=0}} \right) \frac{P_{u,fi,m=0}}{P_{u,a,m=0}} \qquad (9.12)$$

In equation (9.12), P_u is the squash load of the composite cross-section and f_c the concrete cylinder strength. Subscript "a" refers to ambient temperature and "$m = 0$" refers to the reference combination of grades of steel and concrete.

Equation (9.12) expresses a linear relationship as shown in Figure 9.7. This linear relationship is bounded by two points, one at ambient temperature and the other after a long period of fire exposure. At ambient temperature, equation (9.12) gives the ratio of the squash load of a composite column with the design grades of steel and concrete to that using the reference grades. After a prolonged fire exposure when the composite

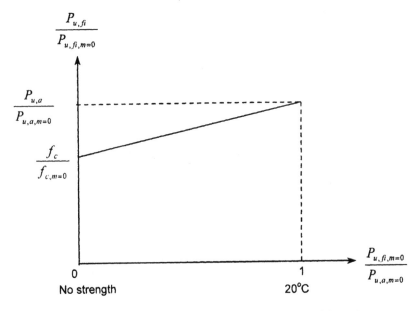

Figure 9.7 Calculation of squash load for unprotected concrete filled columns.

cross-section has lost nearly all its strength, the right hand side of equation (9.12) becomes the ratio of the two concrete strengths ($f_c/f_{c,m=0}$).

The same reasoning for the squash load of a composite cross-section may be applied to the rigidity. However, since the second moment of area is related to the square of the distance from the centre of the composite cross-section, contributions from the steel shell cannot be ignored even at very high temperatures. Also, there is no change in the modulus of elasticity of different grades of steel and the change in the modulus of elasticity of different grades of concrete is small. Therefore, it is reasonable to assume that the ratio of the column rigidity at elevated temperatures to that at ambient temperature is independent of the grades of steel and concrete. Therefore, for the rigidity of a composite cross-section, equation (9.13) is obtained

$$\frac{(EI)_{fi}}{(EI)_{fi,m=0}} = \frac{(EI)_a}{(EI)_{a,m=0}} \tag{9.13}$$

To apply equations (9.12) and (9.13), results should be available for the squash load and rigidity of all steel tubular sizes under all standard fire resistance ratings for the reference combination of grades of steel and concrete. Table 9.5 provides the squash load and rigidity for all concrete

Table 9.5 Reduced squash load and rigidity for different concrete filled CHS sections

Section size		30 min		60 min		90 min		120 min	
		$P_{u,fi}$	$(EI)_{fi}$	$P_{u,fi}$	$(EI)_{fi}$	$P_{u,fi}$	$(EI)_{fi}$	$P_{u,fi}$	$(EI)_{fi}$
168.3*	5.0	0.502	0.368	0.2171	0.1574				
	6.3	0.5002	0.3608	0.1953	0.1432				
	8.0	0.5096	0.3626	0.1744	0.1297				
	10.0	0.5203	0.3739	0.1571	0.1187				
	12.5	0.5721	0.4094	0.1427	0.1095				
193.7*	5.0	0.5451	0.4190	0.2127	0.1946	0.1627	0.1113		
	6.3	0.5419	0.4015	0.2453	0.1762	0.1452	0.1013		
	8.0	0.493	0.4008	0.2192	0.1584	0.1279	0.0917		
	10.0	0.5691	0.4112	0.1974	0.1439	0.1127	0.0835		
	12.5	0.6140	0.4525	0.1787	0.1314				
	16.0	0.6817	0.5155	0.1629	0.1206				
219.1*	4.5	0.5851	0.4545	0.3327	0.2401	0.2222	0.1477	0.1449	0.0926
	5.0	0.5801	0.4469	0.3189	0.2301	0.2120	0.1414	0.1380	0.0889
	6.3	0.5751	0.4353	0.2889	0.2086	0.1895	0.1279	0.1230	0.0808
	8.0	0.5807	0.4320	0.2592	0.1875	0.1668	0.1145	0.1078	0.0729
	10.0	0.6026	0.4468	0.2337	0.1685	0.1466	0.1029		
	12.5	0.6462	0.4866	0.2115	0.1537	0.1279	0.0925		
	16.0	0.7131	0.5484	0.1920	0.1395	0.1095	0.0826		
	20.0	0.7874	0.6164	0.1798	0.1299				
244.5*	5.0	0.6032	0.4776	0.3587	0.2633	0.2601	0.1740	0.1823	0.1149
	6.3	0.6033	0.4646	0.3262	0.2389	0.2335	0.1571	0.1629	0.1039
	8.0	0.6075	0.4596	0.2937	0.2148	0.2062	0.1401	0.1430	0.0928
	10.0	0.6307	0.4773	0.2653	0.1938	0.1817	0.1251	0.1252	0.0830
	12.5	0.6727	0.5153	0.2403	0.1752	0.1587	0.1115	0.1085	0.0741
	16.0	0.7383	0.5756	0.2177	0.1579	0.1357	0.0983		
	20.0	0.8087	0.6359	0.2030	0.1457	0.1175	0.0883		
	25.0	0.8689	0.6584	0.2069	0.1401	0.1022	0.0807		
273.0*	5.0	0.6376	0.5083	0.3970	0.2972	0.3063	0.2081	0.2310	0.1445
	6.3	0.6308	0.4972	0.3627	0.2705	0.2766	0.1881	0.2075	0.1305
	8.0	0.6340	0.4880	0.3278	0.2435	0.2457	0.1678	0.1831	0.1162
	10.0	0.6578	0.5074	0.2970	0.2198	0.2176	0.1497	0.1609	0.1034
	12.5	0.6978	0.5430	0.2692	0.1983	0.1909	0.1329	0.1397	0.0915
	16.0	0.7613	0.6010	0.2437	0.1780	0.1637	0.1161	0.1181	0.0798
	20.0	0.8257	0.6489	0.2285	0.1636	0.1418	0.1034	0.1006	0.0716
	25.0	0.8852	0.6703	0.2329	0.1563	0.1231	0.0932		
323.9*	5.0	0.6793	0.5554	0.4542	0.3503	0.3271	0.2619	0.3091	0.1999
	6.3	0.6710	0.5392	0.4179	0.3206	0.3390	0.2380	0.2804	0.1811
	8.0	0.6741	0.5340	0.3802	0.2899	0.3039	0.2132	0.2499	0.1614
	10.0	0.6953	0.5506	0.3462	0.2623	0.2713	0.1905	0.2216	0.1434
	12.5	0.7323	0.5827	0.3148	0.2366	0.2397	0.1690	0.1941	0.1263
	16.0	0.7927	0.6371	0.2849	0.2116	0.2089	0.1472	0.1654	0.1089
	20.0	0.8484	0.6682	0.2707	0.1944	0.1801	0.1300	0.1416	0.0949
	25.0	0.9068	0.6880	0.2721	0.1834	0.1566	0.1158	0.1204	0.0829
355.6*	6.3	0.6912	0.5623	0.4477	0.3486	0.3708	0.2654	0.3169	0.2096
	8.0	0.6949	0.5585	0.4089	0.3163	0.3341	0.2385	0.2842	0.1875
	10.0	0.7150	0.5738	0.3735	0.2869	0.299	0.2137	0.2535	0.1670

	12.5	0.7506	0.6043	0.3405	0.2592	0.2658	0.1898	0.2233	0.1474
	16.0	0.8090	0.6564	0.3084	0.2317	0.2304	0.1654	0.1915	0.1271
	20.0	0.8604	0.6791	0.2945	0.2129	0.2011	0.1459	0.1649	0.1105
	25.0	0.9178	0.6979	0.2938	0.1996	0.1754	0.1297	0.1408	0.0961
406.4*	6.3	0.7204	0.5964	0.4899	0.3897	0.4160	0.3056	0.3671	0.2520
	8.0	0.7254	0.5952	0.4500	0.3554	0.3775	0.2762	0.3319	0.2267
	10.0	0.7439	0.6085	0.4130	0.3235	0.3409	0.2487	0.2983	0.2030
	12.5	0.7771	0.6362	0.3777	0.2931	0.3045	0.2221	0.2648	0.1798
	16.0	0.8306	0.6813	0.3436	0.2624	0.2659	0.1944	0.2289	0.1554
	20.0	0.8768	0.6953	0.3284	0.2407	0.2335	0.1719	0.1984	0.1351
	25.0	0.9321	0.7123	0.3246	0.2239	0.2045	0.1526	0.1705	0.1171
	32.0	0.9900	0.7352	0.3338	0.2124	0.1798	0.1362	0.1433	0.1004
457.0*	6.3	0.7470	0.6289	0.5252	0.4250	0.4548	0.3422	0.4096	0.2897
	8.0	0.7522	0.6284	0.4847	0.3892	0.4155	0.3112	0.3728	0.2622
	10.0	0.7687	0.6394	0.4467	0.3556	0.3775	0.2819	0.3372	0.2359
	12.5	0.7987	0.6632	0.4099	0.3230	0.3393	0.2527	0.3012	0.2098
	16.0	0.8968	0.6999	0.3747	0.2859	0.2980	0.2220	0.2622	0.1821
	20.0	0.8894	0.7098	0.3582	0.2658	0.2629	0.1966	0.2286	0.1586
	25.0	0.9422	0.7245	0.3518	0.2465	0.2310	0.1745	0.1974	0.1375
	32.0	0.9905	0.7451	0.3616	0.2357	0.2047	0.1550	0.1666	0.1174
	40.0	0.9930	0.7675	0.4159	0.2723	0.1911	0.1423	0.1430	0.1030
508.0*	6.3	0.7709	0.6598	0.5555	0.4563	0.4902	0.3771	0.4458	0.3230
	8.0	0.7746	0.6572	0.5154	0.4200	0.4503	0.3448	0.4082	0.2939
	10.0	0.7888	0.6650	0.4771	0.3855	0.4113	0.3136	0.3714	0.2659
	12.5	0.8161	0.6859	0.4395	0.3515	0.3716	0.2823	0.3338	0.2376
	16.0	0.8587	0.7143	0.4044	0.3168	0.3282	0.2489	0.2924	0.2072
	20.0	0.8990	0.7222	0.3860	0.2906	0.2908	0.2209	0.2563	0.1811
	25.0	0.9497	0.7348	0.3773	0.2688	0.2564	0.1961	0.2225	0.1573
	32.0	0.9905	0.7541	0.3869	0.2579	0.2282	0.1737	0.1887	0.1343
	40.0	0.9929	0.7756	0.4416	0.2957	0.2122	0.1585	0.1625	0.1175
	50.0	0.9948	0.7956	0.5166	0.3509	0.2067	0.1483	0.1427	0.1057
539.0*	20.0	0.9075	0.7334	0.4109	0.3134	0.3164	0.2437	0.2816	0.2021
	25.0	0.9567	0.7444	0.4000	0.2891	0.2804	0.2167	0.2456	0.1760
	32.0	0.9904	0.7625	0.4102	0.2794	0.2503	0.1919	0.2094	0.1507
	40.0	0.9928	0.7832	0.4641	0.3167	0.2320	0.1743	0.1813	0.1322
	50.0	0.9947	0.8065	0.5393	0.3729	0.2259	0.1622	0.1583	0.1185
610.0*	20.0	0.9155	0.7436	0.4332	0.3339	0.3401	0.2655	0.3060	0.2232
	25.0	0.9635	0.7536	0.4208	0.3081	0.3025	0.2363	0.2684	0.1954
	32.0	0.9904	0.7706	0.4328	0.3010	0.2706	0.2090	0.2300	0.1679
	40.0	0.9927	0.7904	0.4849	0.3384	0.2503	0.1893	0.1997	0.1472
	50.0	0.9947	0.8129	0.5598	0.3931	0.2437	0.1756	0.1745	0.1315
660.0*	20.0	0.9232	0.7532	0.4538	0.3532	0.3616	0.2854	0.3285	0.2434
	25.0	0.9700	0.7623	0.4404	0.3260	0.3230	0.2547	0.2892	0.2136
	32.0	0.9903	0.7783	0.4537	0.3213	0.2896	0.2255	0.2489	0.1840
	40.0	0.9927	0.7971	0.5037	0.3548	0.2677	0.2040	0.2167	0.1614
	50.0	0.9946	0.8189	0.5782	0.4116	0.2604	0.1887	0.1898	0.1439

Source: Wang (2000a).

filled Corus manufactured circular steel sections. For this table, concrete is assumed to be grade C30 (Cylinder strength $= 25\,N/mm^2$, Young's modulus $= 32000\,N/mm^2$) and steel grade is S275 (yield strength $= 275\,N/mm^2$, Young's modulus $= 205\,000\,N/mm^2$). All values in Table 9.5 have been obtained by setting the values of φ in equation (9.8) and material partial safety factors to unity. Results in Table 9.5 are expressed as ratios of the squash load and rigidity of a composite column in fire design to those at ambient temperature. An example is provided below to show how Table 9.5 and equations (9.12) and (9.13) are used to directly obtain the squash load and rigidity of a composite cross-section without having to calculate the temperatures or to perform the integrations in equations (9.7) and (9.8).

It must be made clear that values in Table 9.5 have been obtained by using the strength and stiffness degradation factors in EC4 for both steel and concrete. For high strength concrete which has radically different degradation factors, this simplified method cannot be used.

Example 1. *Find the squash load and rigidity for a C60 (Cylinder strength 60 N/mm², Young's modulus 37000 N/mm²) grade concrete filled circular hollow steel section (CHS 168.3 × 5.0) of steel grade S355 for the standard fire resistance rating of 60 min.*

1 Squash load and rigidity at ambient temperature:

For S355 steel and C60 concrete:

$$P_{u,a} = 1894.7\,kN, (EI)_a = 2895\,kN.m^2$$

For S275 steel and C30 concrete:

$$P_{u,a} = 1197.4\,kN, (EI)_a = 2740.9\,kN.m^2$$

2 Squash load and rigidity for 60 min of the standard fire resistance for S275 steel and C30 concrete:

From Table 9.5:

$$P_{u,fi} = 1197.4 \times 0.2171 = 260.0\,kN$$
$$(EI)_{fi} = 2740.9 \times 0.1574 = 431.4\,kN.m^2$$

3 Required squash load and rigidity at 60 min:

Strength from equation (9.12):

$$P_{u,fi}/260.0 = 50/25 + (1894.7/1197.4 - 50/25) \times 0.2171 = 1.901,$$
giving $P_{u,fi} = 496.3\,kN$

Rigidity from equation (9.13):

$$(EI)_{fi}/431.4 = 2895/2740.9 = 1.056, \text{giving } (EI)_{fi} = 455.6\,\text{kN.m}^2$$

So far only unreinforced composite cross-sections have been dealt with. Including contributions from reinforcing bars is relatively straightforward and equations (9.7) and (9.8) may be directly used. Since only a few reinforcing bars will be used, usually located at equal distance from the external surface of the composite cross-section, the reinforcement temperature only needs to be calculated once.

If it is necessary to include material partial safety factors, the above method can still be used. It is a simple matter to treat the material partial safety factors as giving a different combination of grades of steel and concrete.

9.3.2.3   Protected columns with unreinforced concrete

For a protected composite column, an additional design parameter is the steel temperature. For a given steel temperature, the squash load and rigidity of a composite cross-section may be linearly related to the standard fire exposure time. This linear relationship is shown in Figure 9.8. It is

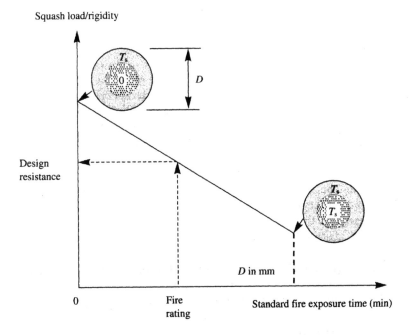

Figure 9.8 Squash load/rigidity of a protected concrete filled column.

bounded by two points. One point is at zero time when it is assumed that the steel is at the design temperature and the concrete is cold. The other point is after a long fire exposure when the concrete is heated to the same temperature as the steel. The minimum time (in minutes) taken to reach uniform temperature distribution in the composite cross-section is numerically equal to the outside dimension of the composite cross-section in mm.

The minimum time–diameter relationship is not entirely coincidental and may be obtained by approximate theoretical consideration. This time should be related to the thermal penetration time of the composite cross-section, which indicates the time for the concrete in the section centre to "feel" the heat transferred from the steel tube. For a flat surfaced concrete slab, the thermal penetration time (t_p) is expressed as (Drysdale 1999)

$$t_p = \frac{\delta^2}{4\alpha} \quad \text{with concrete diffusivity} \quad \alpha = \frac{k}{\rho c} \tag{9.14}$$

where δ is the thickness, k the thermal conductivity, ρ the density and c the specific heat of the concrete slab.

However, for a circular or square section with progressively reducing perimeters, the thermal resistance of each layer starting from the external surface is progressively reduced, the effect of which is equivalent to increasing the concrete thermal conductivity by a linear function related to the overall dimension of the concrete section. Thus, for the problem under consideration, equation (9.14) is now expressed as:

$$t_p \propto \frac{\delta^2}{k} \propto \frac{\delta^2}{\delta} \propto \delta \tag{9.15}$$

Therefore, the minimum time to reach uniform temperature distribution is linearly related to the dimension of the concrete section. By inspection of results of a comprehensive parametric study, Wang (2000a) has found that the linear slope coefficient is near to 1.0 if the time is taken in minutes and the diameter is taken in mm. As the steel temperature is almost uniform, the dimension in Figure 9.8 should be taken as that of the concrete core. However, as the steel tube is comparatively thin and the values of the strength and rigidity of the composite cross-section are not very sensitive to a small change in the minimum time for the cross-section to reach uniform temperature distribution, the external dimension of the steel tube may conveniently be used.

Thus, the squash load ($P_{u,fi}$) of a protected composite cross-section at any given maximum steel temperature (T_s) and fire exposure time (t in minutes) may be calculated using the equation:

$$P_{u,fi} = \frac{t}{D}\left(P_{u,TS=TC=T_s} - P_{u,TS=T_s,TC=0}\right) + P_{u,TS=T_s,TC=0} \tag{9.16}$$

Table 9.6 Squash load and rigidity of a protected composite cross-section[a]

	Composite section temperature condition[b]		
	Ambient temperature	$T_S = 800°C, T_C = 0°C$	$T_S = T_C = 800°C$
$(P_u)_c$	941.1 kN	941.1	$941.1 \times 0.15 = 147.6$
$(P_u)S$	910.6 kN	$910.6 \times 0.11 = 100.2$	100.2
$\Sigma(P_u) =$	1894.7 kN	1084.3	247.8
$(EI)_c$	1140.5 kN.m^2	1140.5	$1140.5 \times 0.15 = 171.1$
$(EI)s$	1754.5 kN.m^2	$1754.5 \times 0.09 = 157.9$	157.9
$\Sigma(EI) =$	2895 kN.m^2	1298.4	329

[a] Subscripts "c" and "s" refer to concrete and steel; P_u and EI represent the squash load and rigidity of the component material.
[b] Bold face indicates additional calculations compared to ambient temperature.

where D is the outside dimension of the steel tube in mm. Subscript "$TS = TC = T_s$" indicates that both steel and concrete of the composite cross-section are at the same temperature T_s. Subscript "$TS = T_s, TC = 0$" indicates that the steel temperature is T_s and the concrete is cold.

Similarly, the rigidity of the composite cross-section is obtained from

$$(EI)_{fi} = \frac{t}{D}\left(EI_{TS=TC=T_s} - EI_{TS=T_s,TC=0}\right) + EI_{TS=T_s,TC=0} \qquad (9.17)$$

An example is provided below to illustrate how this method may be applied.

Example 2. *For the composite cross-section in example 1, assume it is now protected and the steel shell temperature is 800°C. Calculate its squash load and rigidity using equations (9.16) and (9.17).*

Table 9.6 presents the results of these calculations. Additional calculations are highlighted in bold.

Using equations (9.16) and (9.17), the squash load and rigidity of the composite cross-section are:

Squash load:

$$P_{u,fi} = 1084.3 + 60/168.3 \times (247.8 - 1084.3) = 786.1 \, kN$$

Rigidity:

$$(EI)_{fi} = 1298.4 + 60/168.3 \times (329 - 1298.4) = 952.8 \, kN.m^2$$

9.3.2.4   Temperatures in protected steel tubes

As can be seen from equations (9.16) and (9.17), the protected steel tube temperature is the most important design parameter. For hollow steel tubes, equation (6.45) in Chapter 6 may be used. For concrete filled tubes the effect

Table 9.7 Fire protection thickness modification factor

$A_p/V(m^{-1})$	C
50–75	1.00
75	1.00
100	0.92
125	0.88
150	0.81
175	0.75
200	0.69
260–300	0.55

Source: BSI (1990b). Reproduced with the permission of the British Standards Institution under licence number 2001SK/0298.

of concrete acting as a heat sink (which is to reduce the steel temperature) should be included. Equivalently, if the same steel temperature is considered, the fire protection thickness for a concrete filled tube should be less than that for a bare steel tube. This approach has been adopted in Part 8 and Table 9.7 gives modification factors to reduce the fire protection thickness for concrete filled tubes from those obtained for bare steel tubes. Alternatively, equation (6.45) may be used to calculate the steel temperature of a concrete filled tube by using a section factor obtained by multiplying the section factor of the bare steel tube by the values in Table 9.7. Provided the value of Φ in equation (6.45) is small, which is usually the case, the inaccuracy should be small. For intumescent coating protected columns results of steel temperatures from the manufacturers' fire tests should be consulted.

9.3.2.5   Other simple design methods

In Europe, research studies (ECCS 1988) on composite columns in fire have resulted in the development of the design rules in EC4. In North America, much of the design guidance has been developed as a result of extensive fire tests on concrete filled circular and square columns at the National Research Council of Canada (Lie and Chabot 1992; Lie and Kodur 1996). Fire exposure was according to ASTM E119-88 (ASTM 1985) which is very similar to the standard fire exposure of other countries. Test parameters included the steel tube size, the steel wall thickness, the column load level, the column effect length, the concrete strength and the type of aggregate. From the results of these fire tests, regression analyses were carried out and the following equation may be directly used to obtain the standard fire resistance rating of an unprotected concrete filled column (Kodur 1998, 1999):

$$FR = f_1 \frac{(f_c + 20)}{(L_e - 1000)} D^2 \sqrt{\frac{D}{C}} \tag{9.18}$$

where:

FR = fire resistance rating in minutes;

f_c = cylinder strength of concrete in N/mm^2;

L_e = column effective length in mm;

D = outside dimension (diameter or width) of steel tube in mm;

C = applied load in kN;

f_1 = a constant.

It will be noticed that equation (9.18) does not contain any reference to the steel tube thickness. This is based on the observation that contributions from the unprotected steel tube were very small at high temperatures.

Factor f_1 depends on the type of aggregate, the shape of the steel tube; and whether concrete is plain, bar reinforced or fibre reinforced. It depends on the type of aggregate because carbonate aggregates have substantially higher heat capacity than siliceous aggregates, leading to much lower concrete temperatures and higher fire resistance. Values of f_1 can be obtained from Tables 9.8 and 9.9.

Table 9.8 Values of f_1 for plain and fibre reinforced concrete filled columns

Type of aggregate		Plain concrete	Fibre reinforced concrete
CHS	Siliceous	0.07	0.075
	Carbonate	0.08	0.085
SHS	Siliceous	0.06	0.065
	Carbonate	0.07	0.075

Source: Kodur (1999). Reprinted from the *Journal of Constructional Steel Research*, "Performance-based fire resistance design of concrete-filled steel columns". Vol. 51, pp. 21–36, Kodur (1999) with permission from Elsevier Science.

Table 9.9 Values of f_1 for bar reinforced concrete filled columns

Aggregate type	% steel concrete reinforcement	cover thickness (mm)	f_1 for CHS	f_1 for SHS
Siliceous	<3%	<25	0.075	0.065
		≥25	0.08	0.07
	≥3%	<25	0.08	0.07
		≥25	0.085	0.075
Carbonate	<3%	<25	0.085	0.075
		≥25	0.09	0.08
	≥3%	<25	0.09	0.08
		≥25	0.095	0.075

Source: Kodur (1999). Reprinted from the *Journal of Constructional Steel Research*, "Performance-based fire resistance design of concrete-filled steel columns". Vol. 51, pp. 21–36, Kodur (1999) with permission from Elsevier Science.

Example 3. *Using the same data as in Example 1, calculate the fire resistance rating of a composite column with an effective length of 3 m.*

Using results from Example 1 and the column design equations in Section 8.4.1, the column slenderness is $\overline{\lambda}_{fi} = \sqrt{496.3/455.6} = 1.044$, giving a strength reduction factor of $\chi_{fi} = 0.515$ according to equation (8.13). Equation (8.4) gives the load carrying capacity of the composite column at $0.515 \times 496.3/1.2 = 213\,kN$.

Substituting this load into equation (9.18) with $f_1 = 0.07$, the fire resistance rating is found to be 70.5 min. This time is close to the 60 min that was assumed in the previous example. Clearly, this good agreement is coincidental. However, since these two methods have been developed independently, this example gives some confidence in both methods.

9.3.3   Effect of eccentricity

For a composite column, design calculations for combined axial load and bending moment are already complicated at ambient temperature. With the introduction of non-uniform temperature distributions in fire, accurate treatment can only be obtained by sophisticated numerical procedures. Fortunately, the conclusions reached for steel columns (Section 8.4.2 in the previous chapter) with regard to bending moments are also applicable to composite columns. Therefore, it is acceptable to ignore the influence of bending moments transferred by the adjacent beams in continuous composite construction.

Considering only the bending moment from eccentricity, EC4 gives

$$P_{c,fi,\delta} = P_{c,fi} \frac{P_{c,\delta}}{P_c} \tag{9.19}$$

where P_c is the composite column axial strength at ambient temperature; $P_{c,\delta}$ is the reduced column strength due to eccentricity δ at ambient temperature and $P_{c,fi}$ is the column axial strength in fire as calculated in previous sections.

A detailed evaluation shows that the above equation is conservative for small cross-sections and may not be safe for large cross-sections. However, within the realistic range of eccentricity ($\delta <$ section dimension), accuracy of the above equation is acceptable (Wang 2001c).

9.3.4   High strength concrete filled columns

With the introduction of high strength concrete, the load carrying capacity of a concrete filled column can be further enhanced. However, the increase in the fire resistance is relatively small because high strength concrete loses its strength at a much lower temperature than normal strength concrete. By adding a small amount of steel fibre to the concrete, the performance of high strength concrete can be much improved and the performance of fibre reinforced high strength concrete filled steel columns is similar to that with normal strength concrete filling (Kodur and Wang 2001). Provided the strength and stiffness retention factors are available, equations (9.7) and (9.8) can also be used. However, the various simplified methods in the previous section can not be applied.

9.4 CONCLUDING REMARKS

This chapter has introduced the design of composite slabs, beams and columns for fire resistance. For composite slabs and beams, the main problem is related to obtaining their temperature distributions. This chapter has presented some temperature calculation methods. For composite columns, this chapter has described a few simplified calculation methods for easy implementation of the general design equations given in EC4.

Chapter 10

Steel and composite structures without fire protection

The quest for knowledge is one of the main drivers of research to investigate the behaviour of steel and composite structures under fire conditions. However, it should be recognized that the desire to reduce or eliminate fire protection to steelwork is an equally strong incentive to carry out these studies. Fire protection to steelwork can represent a significant part of the total steel structural cost and the elimination of fire protection to steelwork represents a significant saving in construction cost to the client. But more importantly, by reducing the use of fire protection, steel becomes more competitive and the steel industry can benefit from an increased market share. It is not surprising that the steel industry worldwide has been a major supporter and sponsor of research studies of the behaviour of steel structures in fire. It is also appropriate that this book should include a chapter to summarize methods whereby fire protection to steelwork may be eliminated.

The general condition for a satisfactory FR design is that

$$\text{Fire resistance of a structure} \geq \text{required fire resistance} \qquad (10.1)$$

It follows that to use unprotected steel, either the fire resistance of the structure must be increased or the required fire resistance must be reduced or both.

10.1 REDUCING THE FIRE RESISTANCE REQUIREMENT

Fire resistance design has traditionally been based on the standard fire exposure. The minimum required standard fire resistance rating is usually 30 min. As pointed out in Chapter 3, in general, unprotected conventional structures (I beams, H columns etc.) cannot achieve a fire resistance rating of 30 min and fire protection is necessary. Under realistic fire conditions, the fire severity experienced by a steel structure may be much lower. Chapter 7 has provided a detailed description of how to determine more realistic fire exposure conditions. Successful examples of eliminating fire protection to steelwork by applying more realistic fire exposure conditions include steel structures subject to local burning (e.g. car parks) and external steelwork.

Chapter 7 adopts a deterministic approach and the fire exposure is based on the worst-case scenario. In the deterministic approach, a fire will always occur in an enclosure and will grow into the flashover phase. However, in more realistic situations, whether or not a fire will break out in an enclosure is probabilistic in nature and once ignited, the fire may not grow into flashover if fire detection and fire fighting are successful. In other words, fire resistance design should adopt a probabilistic approach. Using the probabilistic approach, the risk of a realistic fire may be much lower than the worst case. One way of utilizing the benefit of the reduced fire risk is to use a reduced fire load in deterministic design calculations. In the so-called Natural Fire Safety Concept (Schleich 1996), the design fire load is obtained from

$$q_{t,d} = \gamma_n \times \gamma_q \times q_{k,f} \tag{10.2}$$

where $q_{k,f}$ is the characteristic fire load; $\gamma_n(<1)$ is a factor accounting for active fire protection measures; and $\gamma_q(>1)$ is a safety factor whose value depends on the consequence of failure and frequency of fires.

Quantitative fire-risk assessment incorporating real life fire statistics should be used to determine these factors. As an approximate guidance, the following values have been suggested within the Natural Fire Safety Concept (Schleich 1996) and Eurocode 1 Part 1.2 (CEN 2000a). The safety factor is given by

$$\gamma_q = \gamma_{q1}\gamma_{q2} \tag{10.3}$$

where γ_{q1} relates to the consequence of structural failure and is a function of the compartment size under fire attack and the number of storeys of a building. Values of γ_{q1} may be taken from Table 10.1.

The safety factor γ_{q2} accounts for the frequency of fires. It is a function of the type of occupancy and its value may be taken from Table 10.2.

Table 10.1 Fire load modification factor γ_{q1} to account for building characteristics

Compartment Floor area A_f (m^2)	Safety factor γ_{q1} for number of storeys			
	$N = 1$	$N = 2$	$2 < N \leq 10$	$N > 10$
≤ 2500	1.0	1.25	1.5	2.0
$2500 < A_f \leq 5000$	1.05	1.4	1.75	2.5
$5000 < A_f \leq 10\,000$	1.1	1.5	N/A	N/A
$10\,000 < A_f \leq 20\,000$	1.2	1.6	N/A	N/A

Source: CEN (2000a). Reproduced with the permission of the British Standards Institution under licence number 2001SK/0298.

Table 10.2 Fire load modification factor γ_{q2} to account for frequency of fire

Safety factor γ_{q2}	Frequency of fire	Examples
0.85	Small	Art gallery, museum
1.0	Normal	Residence, hotel, paper industry
1.2	Mean	Manufactory for machinery & engineering
1.45	High	Chemical laboratory, painting workshop
1.80	Very high	Manufactory of fireworks or paints

Source: CEN (2000a). Reproduced with the permission of the British Standards Institution under licence number 2001SK/0298.

The modification factor γ_n may be broken into a number of items to consider the effects of different types of active fire fighting measures. These active fire fighting measures include automatic water extinguishing system (r_{n1}); independent water supplies (r_{n2}); automatic fire detection & alarm $(r_{n3}$ and $r_{n4})$; automatic alarm transmission to fire brigade (r_{n5}); stand-by fire brigade at work (r_{n6}); off site fire brigade (r_{n7}); safe access routes for fire fighting (r_{n8}) and effective fire fighting devices (r_{n9}). Therefore,

$$\gamma_n = \gamma_{n1}\gamma_{n2}\cdots\gamma_{n9} \leq 1.0 \tag{10.4}$$

Values of the reduction factors r_{n1} to r_{r9} are given in Table 10.3.

Example 1. *Consider a hotel building with five storeys. Assume that the hotel has the following average active fire protection measures: sprinkler installation with automatic fire detection by heat; one independent water supply; smoke detectors; automatic alarm transmission to fire brigade; adequate access routes and fire fighting devices.*

Table 10.3 Fire load modification factor γ_n to account for different fire protection measures

Factor γ_n	Active fire fighting measure
$\gamma_{n1} = 0.6$	Sprinklers
$\gamma_{n2} = 0.7, 0.9$ or 1.0	0/½ Independent water supplies
$\gamma_{n3} = 0.9$	Automatic fire detection and alarm by heat
$\gamma_{n4} = 0.8$	Automatic fire detection and alarm by smoke
$\gamma_{n5} = 0.8$	Automatic alarm transmission to fire brigade
$\gamma_{n6} = 0.6$	Works fire brigade
$\gamma_{n7} = 0.7$	Off site fire brigade
$\gamma_{n8} = 1.0/1.5^*$	Access routes to fire fighting
$\gamma_{n9} = 1.0/1.5^*$	Fire fighting devices
$\gamma_n = 0.1-0.54$	Overall minimum value

*denotes insufficient active measure.

Source: CEN (2000a). Reproduced with the permission of the British Standards Institution under licence number 2001SK/0298.

According to the Approved Documents to the Building Regulations in the United Kingdom (ADB 2000), the required fire resistance rating for a hotel is 60 min. If the hotel is constructed using conventional composite beams and steel columns, fire protection to the steelwork will be necessary.

Now assume each hotel room has dimensions of 6 m wide by 4 m deep by 4 m high, with a window of 6 m by 2 m. The opening factor O (cf. equation (7.31)) is $6 \times 2 \times \sqrt{2}/[2 \times (6 \times 4 + 4 \times 4 + 4 \times 6)] = 0.1326 \, \mathrm{m}^{1/2}$. Assume a design fire load of 100 MJ/m^2 of the total enclosure area.

Consider using equation (10.2). Tables 10.1–10.3 will give $\gamma_n = 0.6 \times 0.9 \times 0.9 \times 0.8 \times 0.8 = 0.31$ and $\gamma_q = 1.5$, giving a reduced fire load of $q_{t,d} = 0.31 \times 1.5 \times 100 = 46.5 \, \mathrm{MJ/m}^2$.

For simplicity, using the equivalent time approach of Eurocode 1 Part 1.2 (equation (7.39)) will give an equivalent standard fire resistance time for the reduced design fire load of:

$$t_{e,q} = q_{f,d} \times k_b \times w_f \times k_c = q_{t,d} \frac{1}{\sqrt{O}} k_b k_c$$

$$= 46.5 \times 0.07 \times \frac{1}{\sqrt{0.1326}} \times 13.7 \times 0.1326 = 16.24 \, \mathrm{min}$$

This equivalent time is short and can be achieved by using unprotected steelwork.

10.2   INCREASING THE FIRE RESISTANCE OF UNPROTECTED STEEL STRUCTURES

Given the design fire resistance requirement, the scope for using unprotected steel may be increased if the fire resistance of an unprotected structure is enhanced. Broadly speaking, increasing the fire resistance of an unprotected steel structure may be approached in two ways: (1) by increasing the fire resistance of individual members; or (2) by utilizing the superior fire performance of whole structures.

10.2.1   Enhancing the fire resistance of individual members

The following methods may be used:

- over design at ambient temperature;
- increase the fire resistance of an unprotected steel structure through the integration of structural load-bearing and fire protection functions; and
- use FR steels.

10.2.1.1 Over design at ambient temperature

The fire resistance of a structural member is directly related to the load ratio. If the structural member is designed to sustain the maximum load at ambient temperature, the load ratio for fire design is about 0.6 and fire protection is almost always necessary if design is based on using current design methods such as those in BS 5950 Part 8 (BSI 1990b) or Eurocodes 3 and 4 Part 1.2 (CEN 2000b, 2001). However if the structural member is over designed (i.e. a much heavier member is used so that its load carrying capacity is much greater than the applied load at ambient temperature), the structural member will have a much lower load ratio and may survive the fire attack without fire protection. However, it is not likely that this approach would be a cost-effective method for pure steel structural members as the increased weight of the steel member designed for no fire protection may far outweigh the cost of fire protection to be applied on a smaller steel member with just enough strength at ambient temperature. The areas where benefit may be gained are to design a composite beam as a pure steel one or to ignore the benefit of semi-rigid connection action at ambient temperature. Of these two, the first method is more likely to be more reliable and easier to implement in practice, especially, on many real projects, composite beams are already being considered only as steel beams at ambient temperature.

10.2.1.2 Integration of structural load bearing and fire protection functions of concrete

It should be appreciated that it is very rare for steelwork to be used alone. Steel is usually used in combination with other materials, in particular with concrete. Concrete is not only a structural material, but also has good thermal insulation properties. Therefore, by combining these two functions of concrete, composite structures may be constructed to give inherently high fire resistance. The systems to be described here have made a special consideration of fire resistance in their design and construction. Many of these structural elements have already been mentioned in Chapters 8 and 9. The following paragraphs will give a short description of their main features, their design and inherent standard fire resistance that they can achieve. This should enable the designer to determine quickly a possible structural load bearing system where the main design concern is to use unprotected steelwork. More detailed information may be found in Bailey and Newman (1998).

It should be pointed out that for most applications in the United Kingdom and in Europe, the required standard fire resistance time is 60 min or 30 min. The following systems have been developed to achieve these standard fire resistance times.

Beams

Three types of construction may be used.

1   Slim floor/Asymmetric beam, shown in Figures 10.1a and 10.1b. In the slim floor construction, a wide plate is welded to the bottom flange of a universal column section and composite floor slabs are supported on the wide plate. In an asymmetric steel beam, the bottom flange is rolled wider than the top flange. Both systems use the same principle to achieve unprotected steelwork: the web of the steel section is protected by the concrete and provides the majority of the bending resistance of the steel beam at elevated temperatures. Only the steel section is assumed to have load carrying capacity, but LTB is prevented by the composite slabs. In general, numerical heat transfer analysis is necessary to obtain temperature distributions in the steel cross-section. For the standard fire exposure, Section 8.3.5 has presented a simplified hand calculation method. Once the temperatures are known, either BS 5950 Part 8 or Eurocode 3 Part 1.2 may be used to calculate the bending moment capacity of the steel beam. Various analyses (Bailey 1999; Ma and Makelainen 1999a) indicate that provided the load ratio does not exceed 0.5, this system can provide 60 min of standard fire resistance.

2   Shelf angle beams, shown in Figure 10.1c. In this system, steel angles are welded to the web of a steel beam and these angles are used to support precast floor units. This system is mainly used to reduce the structural depth of the floor. Since the angles, the upper flange and the upper portion of the web of the steel section are shielded from fire exposure,

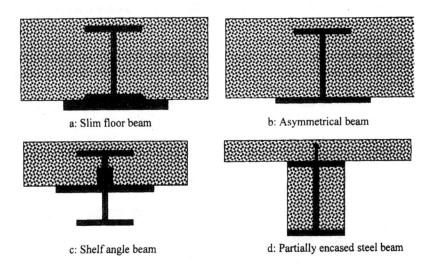

a: Slim floor beam   b: Asymmetrical beam

c: Shelf angle beam   d: Partially encased steel beam

Figure 10.1 Steel/composite beams of high fire resistance.

60 min of fire resistance can be achieved using this system without fire protection. For the standard fire exposure, BS 5950 Part 8 gives a tabulated method to calculate the temperature distributions in a shelf angle beam. These temperatures may be used in combination with the moment capacity method to obtain the fire resistance of the shelf angle beam.

3 Partially encased beams, shown in Figure 10.1d. By casting concrete in between the flanges of a regular universal beam section, only the downward side of the lower flange will be exposed to fire. Both the web and the upper flange are shielded from fire exposure and can provide high structural resistance. Composite floor slabs may be connected to the top of the partially encased steel beam via shear connectors to obtain composite action. Since concrete is cast between flanges of the steel section, no temporary formwork is necessary. By using reinforcement, fire resistance of up to 3 h can be obtained without fire protection to the steelwork. Fire resistant design of this system is given in Eurocode 4 Part 1.2.

Table 10.4 summarizes the standard fire resistance rating that can be achieved by different types of unprotected beams.

Columns

Three types of unprotected columns may be used.

1 Columns with blocked-in webs as shown in Figure 10.2a. In this construction, lightweight aerated concrete blocks are placed between the flanges of a universal steel section. The aerated concrete blocks not only provide good insulation to the column web, they also reduce the average column flange temperature compared to a bare steel column. Fire resistance tests by Corus and Fire Research Station indicate that except for very slender columns, a standard fire resistance rating of 30 min can be achieved without additional fire protection.

2 Partially encased steel columns with unreinforced and reinforced concrete. In a column with blocked-in web, the lightweight aerated concrete

Table 10.4 Standard fire resistance rating of unprotected steel beams

Type of construction	Standard fire resistance time (min)
Bare universal beams	15
Slim floor/Asymmetrical beams	60
Shelf angle beams	60
Partially encased beams	>60

Source: Bailey et al. (1999). Reproduced with the permission of the Institution of Structural Engineers.

only provides insulation to the steel section and the system cannot provide 60-min fire resistance. If normal strength concrete is used to provide composite action, much higher fire resistance can be obtained. If unreinforced concrete is used, 60 min of fire resistance can be obtained. Reinforcement may be used to give much higher fire resistance. Design of this type of column is given in Eurocode 4 Part 1.2.

3 Concrete filled hollow steel sections. Concrete filling of hollow steel sections is a very practical solution to form composite columns. Either unreinforced or reinforced concrete may be used. This type of column has been described in some detail in Chapter 9. To summarize, unreinforced concrete filled columns can achieve 60 min of fire resistance. If reinforcement is used, much higher fire resistance may be obtained. Design of this type of construction may follow Eurocode 4 Part 1.2. However, Chapter 9 has provided a number of alternative design methods that are much easier to use.

To summarize, Table 10.5 gives the standard fire resistance time that can be achieved by different types of unprotected columns.

The usefulness of Tables 10.4 and 10.5 is to enable readers to reach a decision quickly on the form of construction to achieve the required standard fire resistance without fire protection.

a: Blocked-in web   b: Partially encased   c: Concrete filled

Figure 10.2 Steel/composite columns of high fire resistance.

Table 10.5 Standard fire resistance rating of unprotected steel columns

Type of column	Standard fire resistance time (min)
Universal column	15
Blocked-in column	30
Partially encased with unreinforced concrete	60
Partially encased with reinforced concrete	>60
Concrete filled hollow section without reinforcement	60
Concrete filled hollow section with reinforcement	>60

Source: Bailey *et al.* (1999). Reproduced with the permission of the Institution of Structural Engineers.

10.2.1.3 Fire resistant steels

The strength and stiffness of conventional structural steels reduce to about 50% of their original value at a temperature of about 550 °C. It is because of this rapid reduction in the strength and stiffness of steel that fire protection of steel is required in order to reduce the steel temperature. Reductions in the strength of FR steel are slower and unprotected steelwork may be possible.

The best FR steel is perhaps stainless steel. As shown in Figure 8.18, stainless steel retains about 40% of its ambient temperature strength at a very high temperature of about 800 °C. Also stainless steel has a highly reflective surface, which means that the surface emissivity of stainless steel is low, giving low temperatures in unprotected stainless steel in fire. A combination of high strength and low temperature ensures that stainless steel can be used without fire protection. Of course, the high cost of stainless steel means that it can only be used very sparingly.

10.2.2 Utilizing whole building performance in fire

It has long been recognized that the behaviour of the whole building in fire is much better than that of its individual members. By achieving a better understanding of whole building behaviour, it is possible to achieve the objective of eliminating fire protection in conventionally designed and constructed steel-framed buildings. Two methods based on whole building behaviour may be considered.

1 *Utilization of structural redundancy*: Up to now, the design of steel structures for fire safety has been generally based on the assessment of individual structural members, i.e. each structural member should achieve the required fire resistance. However, it should be realized that for a building to remain stable under fire conditions, serving the principal need of containing a fire and preventing its spread, it is not absolutely necessary for every individual structural member to remain stable. Fire protection may be eliminated for some steel members if, in the absence of these members, an alternative load path due to structural redundancy can be developed to retain structural stability.

2 *Developing better understanding of structural behaviour in fire*: Structural fire safety design has been developed based on extending current structural design methods at ambient temperature to elevated temperatures due to the need to avoid large deflections at ambient temperature. The ambient temperature design methods are based on flexural behaviour at small deflections. As has been described in Chapter 5, these current design methods can be very conservative and

structural members may have much higher strength if large deflections are allowed. By exploiting this enhanced load bearing capacity, fire protection may be eliminated.

10.2.2.1 Structural redundancy

In design calculations at ambient temperature, it is assumed that loads acting on a structure are transmitted to the foundation in a pre-determined sequential way. For example, in a multi-storey steel-framed structure, floor loads are transferred by floor slabs to the supporting steel beams, and the beam reactions are then transmitted to the supporting steel columns and thence to the foundations. For fire design, it is usually assumed that the load path selected for the ambient temperature design remains unchanged under fire conditions and that each member in this load path has to have sufficient fire resistance.

It is now appreciated that the load transmission path in a structure is not a fixed one. If one load path breaks down, other alternative load paths may exist and safely transfer the applied loads to the foundation. For example, Figure 10.3a shows part of a steel-framed structure and the usual load carrying path adopted in the ambient temperature design. However, if the floor slabs are designed to have higher load bearing resistance than required at ambient temperature, it is quite possible for the alternative load carrying sequence in Figure 10.3b to develop in fire.

At ambient temperature (Figure 10.3a), the secondary beams are needed to control excessive slab deflections. Under fire conditions, applied floor loads are reduced and large slab deflections are permissible. Thus failure of some secondary beams is permissible provided a sufficient network of beams remains to keep transfer the slab load to the columns. A possible system of this type is shown in Figure 10.3b. In Figure 10.3b, fire protection for the dotted secondary beams is not required. Of course, the design load carrying capacity of the slab can be further increased even to bypass some main steel beams.

Example 2. *Consider the composite floor slab used in the Cardington structure and refer to calculations in Section 5.6.2.*

If the mesh reinforcement is doubled (e.g. by reducing the spacing by half), yield line analysis would give a slab load carrying capacity in fire of about $2 \times 1.95 = 3.9 \, \text{kN/m}^2$. The bending strength of the unprotected secondary steel beam at $900\,^\circ\text{C}$ is $1.14 \, \text{kN/m}^2$, giving a total load carrying capacity of the floor slab system of $5.04 \, \text{kN/m}^2$. This is only slightly lower than the applied load of $5.24 \, \text{kN/m}^2$. It would not be too difficult to find sufficient strength reserves in the floor system to increase its load bearing capacity to $5.24 \, \text{kN/m}^2$, for example by exploring possible enhancements to the strength of the slab by the decking and to the secondary steel beam by semi-rigid connections.

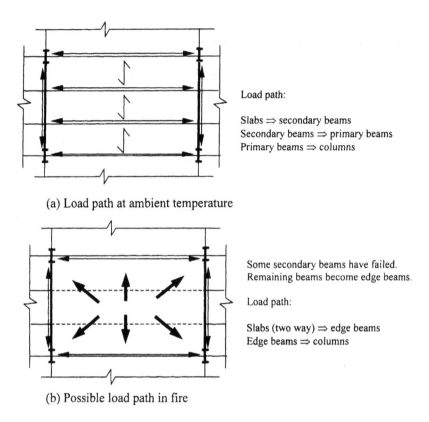

Load path:

Slabs ⇒ secondary beams
Secondary beams ⇒ primary beams
Primary beams ⇒ columns

(a) Load path at ambient temperature

Some secondary beams have failed.
Remaining beams become edge beams.

Load path:

Slabs (two way) ⇒ edge beams
Edge beams ⇒ columns

(b) Possible load path in fire

Figure 10.3 An example of alternative load paths in a structure.

10.2.2.2 Better understanding of structural behaviour at large deflections

Current fire safety design of steel structural members is based on extending current design methods at ambient temperature to elevated temperatures. Since the ambient temperature applications usually involve small deflections, current fire safety design methods are based on the well-understood principles of structural mechanics in flexural bending at small deflections. However, the behaviour of a structure at large deflections can be drastically different and the structure can usually develop substantially higher load carrying capacities. Chapter 5 has described in some detail the mechanisms of catenary action in beams and tensile membrane action in slabs at large deflections. The exploitation of these advanced understandings of structural behaviour is likely to provide the biggest opportunity to use unprotected steel in buildings under fire conditions.

10.3   CONCLUDING REMARKS

Fire engineering of steel and composite structures has come a long way, in particular, the last ten years or so have seen tremendous advances. It is the present writer's belief that within the next few years, research studies will have advanced sufficiently to make it possible to eliminate fire protection to all steel beams in buildings without compromising structural safety in fire. This will require the designers to adopt new understandings of the behaviour of steel structures under fire conditions. Indeed, the purpose of this book is to help achieve these new understandings.

References

Aasen, B. (1985) *An experimental study on steel columns behaviour at elevated temperatures*, The Norwegian Institute of Technology, Trondheim, Norway.

Ala-Outinen, T. and Myllymaki, J. (1995) "The local buckling of RHS members at elevated temperatures", *VTT Research Notes* 1672, Technical Research Centre of Finland.

Alfawakhiri, F. and Sultan, M. A. (2000) "Fire resistance of loadbearing LSF assemblies", *Proceedings of the 15th International Speciality Conference on Cold-Formed Steel Structures*, St. Louis, October.

Al-Jabri, K. S. (1999) "The behaviour of steel and composite beam-to-coloumn connections in fire", Ph.D. thesis, University of Sheffield.

Al-Jabri, K. S., Lennon, T., Burgess, I. W. and Plank, R. J. (1998) "Behaviour of steel and composite beam-column connections in fire", *Journal of Constructional Steel Research* **46**(1–3): 308–309.

Allam, A. M., Fahad, M. K., Liu, T. C. H., Burgess, I. W., Plank, R. J. and Davies, J. M. (1999) "Effects of restraint on the behaviour of steel frames in fire", *Proceedings of the Eurosteel Conference*, CVUT Praha, Czech Republic.

Amdahl, J., Eberg, E. and Holmaas, T. (1995) "Ultimate collapse of offshore structures exposed to fire", *Proceedings of OMAE, Safety and Reliability*, ASME.

Amdahl, J. and Hellan, O. (1992) "Progressive collapse analysis of steel and aluminum structures subjected to fire loads", in J. L. Clarke, F. K. Garas and A. S. T. Armer (eds), *Structures Design for Hazardous Loads*, London, The Institution of Structural Engineers.

American Society for Testing and Materials (ASTM) (1985) *ASTM E 119–83, Standard methods of fire tests of building construction and materials*, ASTM, Philadelphia.

Anderberg, Y. (1983) *Properties of materials at high temperatures, behaviour of steel at high temperatures*, RILEM 44-PHT, Division of Building Fire Safety and Technology, Lund, Sweden.

Anderberg, Y. and Thelandersson, S. (1976) *Stress and deformation characteristics of concrete at high temperatures, 2: Experimental investigation and material behaviour model*, Division of Structural Mechanics and Construction, Lund Institute of Technology.

Anon. (1986) *Fire behaviour of steel and composite construction*, Verlag, TUV, Rheinland.

Approved Document B (ADB) (2000) *Fire Safety*, London: HMSO.

Aribert, J. M. and Randrianstara, C. (1980) "Etude du flambement a des temperatures d'incendie, Action du fluage", *Construction Metallique*, CTICM **4**: 3–22.

Armer, G. S. T. and Moore, D. B. (1994) "Full scale testing on complete multi-storey structures", *The Structural Engineer* **72**(2): 30–32.

Azpiau, W. and Unanue, J. A. (1993) "Buckling curves for hot rolled H profiles submitted to fire", *Report* 97.798-0-ME/V, Labein, Bilbao, Spain.

Babrauskas, V. (1981) "A closed-form approximation for post-flashover compartment fire temperatures", *Fire Safety Journal* **4**: 63–73.

Baddoo, N. and Burgan, B. A. (1998) "Fire resistant design of austenitic structural stainless steel", *Journal of Constructional Steel Research* **46**(1:3): paper No. 243.

Baddoo, N. (1999) "Stainless steel in fire", *The Structural Engineer* **77**(19): 16–17.

Bailey, C. G. (1995) "Simulation of the structural behaviour of steel-framed buildings in fire", Ph.D. thesis, Department of Civil and Structural Engineering, University of Sheffield, UK.

Bailey, C. G., Burgess, I. W. and Plank, R. J. (1996a) "The lateral-torsional buckling of unrestrained steel beams in fire", *Journal of Constructional Steel Research* **36**(2): 101–119.

Bailey, C. G., Burgess, I. W. and Plank, R. J. (1996b) "Analysis of the effects of cooling and fire spread on steel-framed buildings", *Fire Safety Journal* **26**: 273–293.

Bailey, C. G., Burgess, I. W. and Plank, R. J. (1996c) "Computer simulation of a full-scale structural fire test", *The Structural Engineer* **74**(6): 93–100.

Bailey, C. G. (1998a) "Development of computer software to simulate the structural behaviour of steel-framed buildings in fire", *Computers and Structures* **67**: 421–438.

Bailey, C. G. (1998b) "Enhancement of fire resistance of beams by beam-to-column connections", *New Steel Construction* **4**: 44.

Bailey, C. G. (1998c) "Computer modeling of the corner compartment fire test on the large-scale Cardington test frame", *Journal of Constructional Steel Research* **48**: 27–45.

Bailey, C. G. and Newman, G. N. (1998) "The design of steel framed buildings without applied fire protection", *The Structural Engineer* **76**(5): 77–81.

Bailey, C. G. (1999) "The behaviour of asymmetric slim floor steel beams in fire", *Journal of Constructional Steel Research* **50**: 235–257.

Bailey, C. G., Moore, D. B. and Lennon, T. (1999a) "The structural behaviour of steel columns during a compartment fire in a multi-storey braced steel-frame", *Journal of Constructional Steel Research* **52**: 137–157.

Bailey, C. G., Lennon, T. and Moore, D. B. (1999b) "The behaviour of full-scale steel-framed building subjected to compartment fires", *The Structural Engineers* **77**(8): 15–21.

Bailey, C. G. (2000a) "The influence of the thermal expansion of beams on the structural behaviour of columns in steel-framed structures during a fire", *Engineering Structures* **22**: 755–768.

Bailey, C. G. (2000b) "Effective lengths of concrete-filled steel square hollow sections in fire", *Proceedings of the Institution of Civil Engineers, Structures & Buildings* **140**: 169–178.

Bailey, C. G. and Moore, D. B. (2000a) "The structural behaviour of steel frames with composite floor slabs subject to fire, Part 1: Theory", *The Structural Engineer* **78**(11): 19–27.

Bailey, C. G. and Moore, D. B. (2000b) "The structural behaviour of steel frames with composite floor slabs subject to fire, Part 2: Design", *The Structural Engineer* **78**(11): 28–33.

Bailey, C. G. (2001) "Membrane action of unrestrained lightly reinforced concrete slabs at large displacements", *Engineering Structures* 23: 470–483.

Bailey, C. G., White, D. S. and Moore, D. B. (2000) "The tensile membrane action of unrestrained composite slabs simulated under fire conditions", *Engineering Structures* 22: 1583–1595.

Bathe, K. (1996) *Finite Element Procedures*, New Jersey: Prentice-Hall.

Black, M. S. (1973) "Ultimate strength study of two-way concrete slabs", *ASCE Journal of the Structural Division* 101(ST1): 311–324.

Bond, G. V. L. (1975) *Fire and Steel Construction: Water Cooled Hollow Columns*, London: Constrado.

Both, K., Stark, J. W. B. and Twilt, L. (1992) "Thermal shielding near intermediate supports of continuous span composite slabs", *Proceedings of 11th International Speciality Conference on Cold-Formed Steel Structures*, Missouri, October.

Both, C., Stark, J. W. B. and Twilt, L. (1993) "Partial fire protection of composite slabs", *Journal of Constructional Steel Research* 27: 143–158.

Both, C. and Twilt, L. (1994) "3-D analysis of fire-exposed composite slabs", *Proceedings of First International Diana Conference on Computational Mechanics*, Delft, Holland, October.

Both, K., Twilt, L. and Stark, J. W. B. (1995) "Development of European calculation rules for fire exposed composite steel/concrete slabs", *Proceedings of ASIA-FLAM'95*, Hong Kong, March.

Both, C., van Foeken, R. J. and Twilt, L. (1996) "Analytical aspects of the Cardington fire test programme", *Proceedings of the Second Cardington conference*, Building Research Establishment, UK.

Both, C., Fellinger, J. H. H. and Twilt, L. (1997) "Shallow floor construction with deep composite deck: from fire tests to simple calculation rules", *HERON* 42(3): 145–158.

Bressington, P. (1997) "Integrated active fire protection: two systems for the price of one", *Building Engineer*, June, pp. 10–13.

British Standards Institution (BSI) (1985) *British Standard BS 5400, Steel, Concrete and Composite Bridges, Part 5: Code of Practice for Design of Composite Bridges*, British Standards Institution, London.

British Standards Institution (BSI) (1987a) *British Standard 476, Fire Tests on Building Materials and Structures, Part 20: Method for Determination of the Fire Resistance of Elements of Construction (General Principles)*, British Standards Institution, London.

British Standards Institution (BSI) (1987b) *British Standard BS 5950, Structural Use of Steelwork in Buildings, Part 5: Code of Practice for Design of Cold Formed Sections*, British Standards Institution, London.

British Standards Institution (BSI) (1990a) *British Standard BS 5950, Structural Use of Steelwork in Buildings, Part 1: Code of Practice for Design in Simple and Continuous Construction: Hot Rolled Sections*, British Standards Institution, London.

British Standards Institution (BSI) (1990b) *British Standard BS 5950, Structural Use of Steelwork in Buildings, Part 8: Code of Practice for Fire Resistant Design*, British Standards Institution, London.

British Standards Institution (BSI) (2001) *Draft BS 9999, Code of Practice for Fire Safety in the Design, Construction and Use of Buildings*, British Standards Institution, London.

Buchanan, A. H. (2001) *Structural design for fire safety*, Chichester: John Wiley & Sons, Ltd.

Burgess, I. W., El-Rimawi, J. A. and Plank, R. J. (1990) "Analysis of beams with non-uniform temperature profile", *Journal of Constructional Steel Research* **16**: 169–192.

Burgess, I. W., El-Rimawi, J. A. and Plank, R. J. (1991) "Studies of the behaviour of steel beams in fire", *Journal of Constructional Steel Research* **19**: 285–312.

Burgess, I. W., Olawale, A. O. and Plank, R. J. (1992) "Failure of steel columns in fire", *Fire Safety Journal* **18**(2): 183–201.

Burgess, I. W. and Plank, R. J. (1998) "Modelling the fire tests on the Cardington Full-Scale Frame", *Proceedings of the Third Cardington Conference*, Bedfordshire, England.

Carslaw, H. S. and Jaeger, J. C. (1959) *Conduction of heat in solids*, 2nd edn, Oxford: Oxford University Press.

Chan, S. L. and Chan, B. H. M. (2001) "Refined plastic hinge analysis of steel frames under fire", *Steel and Composite Structures* **1**(1): 111–130.

Commission of the European Community (CEC) (1987) "Computer assisted analysis of the fire resistance of steel and composite concrete-steel structures (REFAO-CAFIR)", Final Report, CEC Agreement 7210-SA/502.

Commission of the European Community (CEC) (1997a) "Development of design rules for steel structures subjected to natural fires in large compartments", Draft Final Report, CEC Agreement 7210-SA/210/317/517/619/932.

Commission of the European Community (CEC) (1997b) "Development of design rules for steel structures subjected to natural fires in closed car parks", Draft Final Report, CEC Agreement 7210-SA/211/318/518/620/933.

Connolly, R. J., Wang, Y. C. and Morris, W. A. (1995) "A review of relationships used to predict the burning rate of fully developed fires for application to large compartments", Client Report, CR170/95, Building Research Establishment.

Conseil International du Batiment (CIB W14) (1983) "A conceptual approach towards a probability based design guide on structural fire safety", *Fire Safety Journal* **6**: 24–79.

Conseil International du Batiment (CIB W14) (1986) "Design guide – structural fire safety", *Fire Safety Journal* **10**: 79–137.

Cooke, G. M. E. and Latham, D. J. (1987) *The inherent fire resistance of a loaded steel framework*, Fire Research Station paper PD 57/87, Building Research Establishment.

Cooke, G. M. E., Lawson, R. M. and Newman, G. M. (1988) "Fire resistance of composite deck slabs", *The Structural Engineer* **66**(16): 253–261.

Correia Rodrigues, J. P., Cabrita Neves, I. and Valente, J. C. (2000) "Experimental research on the critical temperature of compressed steel elements with restrained thermal elongation", *Fire Safety Journal* **35**: 77–98.

Couchman, G. H. (1997) "Design of semi-continuous braced frames", *SCI Publication* **183**: The Steel Construction Institute.

Cox, G. (ed.) (1995) *Computational fire modelling*, London, Academic Press.

Crisfield, M. A. (1991) *Nonlinear finite element analysis of solids and structures*, New York, Wiley.

Davis, Langdon & Everest (DLE) (1997) *Spon's Architects' and Builders' Price Book*, London: E & FN Spon.

Dorn, J. E. (1954) "Some fundamental experiments on high temperature creep", *Journal of the Mechanics and Physics of Solids* **3**: 85.

Drysdale, D. (1999) *An Introduction to Fire Dynamics*, 2nd edn, Chichester: John Wiley & Sons.

Dvorkin, E. N. and Bathe, K. J. (1984) "A continuum mechanics based four-node shell element for general nonlinear analysis", *Engineering Computing* **1**: 77–88.

Edwards, M. (1998a) "Reinstatement of concrete filled structural hollow section columns after short duration fires, Phase 2: Standard fire tests on full size columns", *Tubes Division Development Project* No. S2632, British Steel.

Edwards, M. (1998b) "The performance in fire of concrete filled SHS columns protected by intumescent paint", *Proceedings of the Nordic steel construction conference 98*, Bergen, Norway, September, pp. 27–38.

Edwards, M. (2001) "The performance in fire of fully utilized concrete filled SHS columns with external fire protection", in R. Puthli and S. Herion (eds), *Tubular Structures IX*, Lisse: Balkema Publishers, pp. 461–465.

Elghazouli, A. Y. and Izzuddin, B. A. (2000) "Response of idealised composite beam-slab system under fire conditions", *Journal of Constructional Steel Research* **56**: 199–224.

Elghazouli, A. Y., Izzuddin, B. A. and Richardson, A. J. (2000) "Numerical modelling of the structural fire behaviour of composite buildings", *Fire Safety Journal* **35**: 279–297.

El-Rimawi, J. A., Burgess, I. W. and Plank, R. J. (1997) "The influence of connection stiffness on the behaviour of steel beams in fire", *Journal of Constructional Steel Research* **43**(1–3): 1–15.

El-Zanaty, M. H., Murray, D. W. and Bjorhovde, R. (1980) "Inelastic behaviour of multistorey steel frames", *Structural Engineering Report No.* 83, Department of Civil Engineering, University of Alberta.

El-Zanaty, M. H. and Murray, D. W. (1980) Finite element programs for frame analysis, *Strutural Engineering Report No.* 84, Department of Civil Engineering, University of Alberta.

El-Zanaty, M. H. and Murray, D. W. (1983) "Non-linear finite element analysis of steel frames", *ASCE Journal of Structural Division* **109**(ST2): 353–368.

European Convention for Constructional Steelwork (ECCS), (1988) "Calculation of the fire resistance of centrally loaded composite steel-concrete columns exposed to the standard fire", *Technical Note* No. 55, ECCS Technical Committee 3.

European Committee for Standardisation (CEN) (1992a) *DD ENV 1993-1-1, Eurocode 3: Design of Steel Structures, Part 1.1: General Rules and Rules for Buildings*, British Standards Institution, London.

European Committee for Standardisation (CEN) (1992b) *Eurocode 4: Design of Composite Steel and Concrete Structures, Part 1.1: General Rules and Rules for Buildings*, British Standards Institution, London.

European Committee for Standardisation (CEN) (1994) *Eurocode 3: Design of Steel Structures, Part 1.1 Revised Annex J: Joints and Building Frames*, British Standards Institution, London.

European Committee for Standardisation (CEN) (1996) *Eurocode 3: Design of Steel Structures, Part 1.3: Supplementary Rules for Cole-Formed Thin Gauge Members and Sheeting*, British Standards Institution, London.

European Committee for Standardisation (CEN) (2000a) *Draft prEN 1991-1-2, Eurocode 1: Basis of Design and Actions on Structures, Part 1.2: Actions on Structures – Actions on Structures Exposed to Fire*, British Standards Institution, London.

European Committee for Standardisation (CEN) (2000b) *Draft prEN 1993-1-2, Eurocode 3: Design of Steel Structures, Part 1.2: General Rules, Structural Fire Design*, British Standards Institution, London.

European Committee for Standardisation (CEN) (2001) *prEN 1994-1-2, Eurocode 4: Design of Composite Steel and Concrete Structures, Part 1.2: Structural Fire Design*, British Standards Institution, London.

Fahad, M. K., Liu, T. C. H. and Davies, J. M. (2000) "The behaviour of steel beams in fire and the effects of restraint by columns", *Proceedings of the International Conference on Steel Structures of the 2000s*, ECCS/CECM/EKS, Istanbul, Turkey.

Fahad, M. K., Liu, T. C. H., Davies, J. M., Allam, A. M., Burgess, I. W. and Plank, R. J. (2001) "Behaviour of axially restrained steel beams in fire – experimental results", *Journal of Constructional Steel Research*, in press.

Fellinger, J. H. H. and Twilt, L. (1996) "Fire resistance of slim floor beams", *Proceedings of the 2nd ASCE Conference*, Irsee, Germany, June.

Feng, M. Q., Wang, Y. C. and Davies, J. M. (2001) "Behaviour of short thin-walled channel columns at elevated temperatures", *Proceedings of the International Seminar on Steel Structures in Fire*, Shanghai, December.

Forsen, N. E. (1982) "A theoretical study on the fire resistance of concrete structures – CONFIRE", *SINTEF Report* No. STF65 A82062, The Norwegian Institute of Technology.

Forsen, N. E. (1983) "STEELFIRE – Finite element program for nonlinear analysis of steel frames exposed to fire", Multiconsult AS Consulting Engineers, Norway.

Franssen, J. M. (1987) "A study of the behaviour of composite steel-concrete structures in fire", Ph.D. thesis, University of Liege, Belgium.

Franssen, J. M., Schleich, J. B., Talamona, D., Twilt, L. and Both, K. (1994) "A comparison between fire structural fire codes applied to steel elements", *Proceedings of the Fourth International Symposium on Fire Safety Science*, Ottawa, Canada, pp. 1125–1136.

Franssen, J. M., Schleich, J. B. and Cajot, L. G. (1995) "A simple model for the fire resistance of axially loaded members according to Eurocode 3", *Journal of Constructional Steel Research* **35**: 49–69.

Franssen, J. M., Talamona, D., Kruppa, J. and Cajot, L. G. (1998) "Stability of steel columns in case of fire: experimental evaluation", *ASCE Journal of Structural Engineering* **124**(2): 158–163.

Franssen, J. M. (1998) "Improvement of the parametric fire of Eurocode 1 based on experimental test results", *Proceedings of the 6th International Symposium on Fire Safety Science*, France, pp. 927–938.

Franssen, J. M., Kodur, V. K. R. and Mason, J. (2000) *User's manual for SAFIR2001, A computer program for analysis of structures submitted to the fire*, University of Liege, Belgium.

Franssen, J. M. (2000) "Failure temperature of a system comprising a restrained column submitted to fire", *Fire Safety Journal* **34**: 191–207.

Galambos, T. V. (1998) *Guide to Stability Design Criteria for Metal Structures*, 5th edn, New York: John Wiley & Sons, Inc.

Gerlich, J. T. (1995) "Design of loadbearing light steel frame walls for fire resistance", *Fire Engineering Research Report* 95/3, School of Engineering, University of Canterbury.

Gibbons, C., Nethercot, D. A., Kirby, P. A. and Wang, Y. C. (1993) "An appraisal of partially restrained column behaviour in non-sway steel frames", *Proceedings of the Institution of Civil Engineers, Structures and Buildings* 99: 15–28.

Gillie, M. (1999) "Development of generalized stress strain relationships for the concrete slab in shell models", PIT Project Research Report SS1, The University of Edinburgh.

Gillie, M., Usmani, A. and Rotter, M. (2000) "Modelling of heated composite floor slabs with reference to the Cardington experiments", University of Edinburgh.

Gillie, M., Usmani, A. S. and Rotter, J. M. (2001) "A structural analysis of the first Cardington test", *Journal of Constructional Steel Research* 58: 581–601.

Hagen, E. and Haksever, A. (1986) "A contribution for the investigation of natural fires in large compartments", *Proceedings of the First International Symposium on Fire Safety Science*, pp. 149–158.

Hagen, E. (1991) "Zur prognose des Gefahodungspotentials von ramm branden", Ph.D. thesis, University of Braunschweig, Germany.

Harmathy, T. Z. (1967) "A comprehensive creep model", *Journal of Basis Engineering Transactions*, ASME, 396–502.

Harmathy, T. Z. (1972) "A new look at compartment fires, Parts I and II", *Fire Technology* 8: 196–217, 326–351.

Harmathy, T. Z. (1980) "The possibility of characterizing the severity of fires by a single parameter", *Fire and Materials* 4(2): 71–76.

Harmathy, T. Z. (1982) "Normalized heat load: A key parameter in fire safety design", *Fire and Materials* 6(1): 27–31.

Harmathy, T. Z. (1987) "On the equivalent fire exposure", *Fire and Materials* 11: 95–104.

Harmathy, T. Z. and Mehaffey, J. R. (1987) "The normalized heat load concept and its use", *Fire Safety Journal* 12: 75–81.

Hasemi, Y. (1999) "Diffusion flame modeling as a basis for the rational fire safety design of built environment", *Proceedings of the Sixth International Symposium on Fire Safety Science*, France, pp. 3–21.

Hasemi, Y., Yokobayashi, Y., Wakamatsu, T. and Ptchelintsev, A. (1997) "Modelling of heating mechanism and thermal response of structural components exposed to localized fires: a new application of diffusion flame modeling to fire safety engineering", *Proceedings of the Thirteenth Meeting of the UJNR Panel on Fire Research and Safety*, National Institute of Standards and Technology (NIST), March.

Hass, R., Ameler, J., Zies, H. and Lorenz, H. (2001) "Fire resistance of hollow section composite columns with high strength concrete filling", in R. Puthli and S. Herion (eds), *Tubular Structures IX*, Lisse: Balkema Publishers, 467–474.

Hawkins, N. M. and Mitchell, D. (1979) "Progressive collapse of flat plate structures", *ACI Journal* 76(7): 775–808.

Hayes, B. (1968a) "Allowing for membrane action in the plastic analysis of rectangular reinforced concrete slabs", *Magazine of Concrete Research* 20(65): 205–212.

– also Corrigenda and discussion with Morley, C. T. (1969) *Magazine of Concrete Research* **21**(69): 235–236.

Hayes, B. (1968b) "A Study of the Design of Reinforced Concrete Slabs", Ph.D. thesis, University of Manchester.

Hayes, B. and Taylor, R. (1969) "Some tests on reinforced concrete beam-slab panels", *Magazine of Concrete Research* **21**(67): 113–120.

Hinkley, P. L. (1986) "Rates of production of hot gases in roof venting experiments", *Fire Safety Journal* **10**: 57–65.

Hirota, M., Shinoda, K., Nakamura, K. and Kawagoe, K. (1984) "Experimental study on structural behaviour of steel frames in building fire", *Fire Science and Technology* **4**(2): 151–162.

Hosser, D., Dorn, T. and El-Nesr, O. (1994) "Experimental and numerical studies of composite beams exposed to fire", *ASCE Journal of Structural Engineering* **20**(10): 2871–2892.

Hu, B. L., Wang, Y. C. and Davies, J. M. (2001) "Behaviour of restrained columns in fire, experimental results from phase 1 study", *Proceedings of the International Seminar on Steel Structures in Fire*, Shanghai, November.

Huang, Z. H., Burgess, I. W. and Plank, R. J. (1999a) "Nonlinear analysis of reinforced concrete slabs subjected to fire", *ACI Structural Journal* **96**(1): 127–135.

Huang, Z. H., Burgess, I. W. and Plank, R. J. (1999b) "The influence of shear connectors on the behaviour of composite steel-framed buildings in fire", *Journal of Constructional Steel Research* **51**: 219–237.

Huang, Z. H., Burgess, I. W. and Plank, R. J. (1999c) "Three-dimensional modeling of two full-scale fire tests on a composite building", *Proceedings of the Institution of Civil Engineers, Structures and Building* **134**: 243–255.

Huang, Z. H., Burgess, I. W. and Plank, R. J. (2000a) "Effective stiffness modelling of composite concrete slabs in fire", *Engineering Structures* **22**: 1133–1144.

Huang, Z. H., Burgess, I. W. and Plank, R. J. (2000b) "Non-linear modeling of three full-scale structural fire tests", *Proceedings of the First International Workshop on Structures in Fire*, Copenhagen, Denmark, June.

Iding, R., Bresler, B. and Nizamuddin, D. (1977a) "FIRES-T3: A computer program for the fire response of structures – thermal", *Fire Research Group Report* No. UCB FRG 77–15, University of California, Berkeley.

Iding, R., Bresler, B. and Nizamuddin, D. (1977b) "FIRES-RCII, A computer program for the fire response of structures – reinforced concrete frames", *Fire Research Group Report* No. UCB FRG 77–8, University of California, Berkeley.

Institute of Fire Safety Design (IFSD) (1986) TEMPCALC User Manual, IFSD, LUND, IDEON, Sweden.

The Institution of Structural Engineers (ISE) (1996) *Appraisal of Existing Structures*, London: The Institution of Structural Engineers.

International Standards Organization (ISO) (1975) *ISO 834: Fire Resistance Tests, Elements of Building Construction*, Geneva: International Organization for Standardization.

Izzuddin, B. A. (1991) "Nonlinear dynamic analysis of framed structures", Ph.D. thesis, Department of Civil Engineering, Imperial College, London.

Izzuddin, B. A. (1996) "Quartic formulation for elastic beam-columns subject to thermal effects", *ASCE Journal of Engineering Mechanics* **122**(9): 861–871.

Izzuddin, B. A. and Elghazouli, A. Y. (1999) "A simplified model for beam-slab systems subject to fire", *Proceedings of the Eurosteel conference 99*, CVUT Praha, Czech Republic.

Izzuddin, B. A., Song, L., Elnashai, A. S. and Dowling, P. J. (2000) "An integrated adaptive environment for fire and explosion analysis of steel frames – Part 2: verification and application", *Journal of Constructional Steel Research* **53**: 87–111.

Jayarupalingam, N. (1996) "Steel, steel/concrete composite and reinforced concrete beams and columns exposed to fire", Ph.D. thesis, City University.

Jayarupalingam, N. and Virdi, K. S. (1992) "Steel beams and columns exposed to fire hazard", in J. L. Clarke, F. K. Garas and G. S. T. Armer (eds), *Structural Design for Hazardous Loads, The Role of Physical Testing*, London: E & FN Spon, pp. 429–438.

Jeanes, D. C. (1985) "Application of the computer in modeling fire endurance of structural steel floor systems", *Fire Safety Journal* **9**(1): 119–135.

Johansen, K. W. (1962) *Yield Line Theory*, Cement and Concrete Association, London.

Kaneka, H. (1990) "Etude par la méthode des éléments finis du comportement mécanique d'éléments plaques en acier soumis à l'incendie", *Construction Métallique* **1**.

Karlsson, B. and Quintiere, J. G. (2000) *Enclosure Fire Dynamics*, Florida, CRC.

Kasami, H. (1975) "Properties of concrete exposed to sustained elevated temperatures", *Proceedings of 3rd International Conference on Structural Mechanics in Reactor Technology*, London.

Kay, T., Kirby, B. R., Preston, R. R. (1996) "Calculation of the heating rate of an unprotected steel member in a standard fire resistance test", *Fire Safety Journal* **26**(4): 327–350.

Kawagoe, K. (1958) "Fire behaviour in rooms", *Report No. 27*, Building Research Institute, Japan.

Kelly, F. S. and Sha, W. (1999) "A comparison of the mechanical properties of fire-resistant and S275 structural steels", *Journal of Constructional Steel Research* **50**: 223–233.

Kemp, K. O. (1967) "Yield of a square reinforced concrete slab on simple supports, allowing for membrane forces", *The Structural Engineer* **45**(7): 235–240.

Khoury, G. A. (1983) "Transient Thermal Creep of Nuclear Reactor Pressure Vessel Type Concretes", Ph.D. thesis, Department of Civil Engineering, Imperial College of Science, Technology and Medicine.

Khoury, G. A., Grainger, B. N. and Sullivan, P. J. E. (1985) "Transient thermal strain of concrete during first heating cycle to 600 °C", *Magazine of Concrete Research* **37**(133): 195–215.

Khoury, G. A. (1992) "Compressive strength of concrete at high temperatures: a reassessment", *Magazine of Concrete Research* **44**(161): 291–309.

Kimura, M., Ohta, H., Kaneko, H., Kodaira, A. and Fujinaka, H. (1990) "Fire resistance of concrete-filled square steel tubular columns subjected to combined loads", *Takenaka Technical Research Report* **43**: 47–54.

Kirby, B. R., Lapwood, D. G. and Thomson (1986) "The reinstatement of fire damaged steel and iron framed structures", British Steel Corporation, Swinden Laboratories.

Kirby, B. R. and Preston, R. R. (1988) "High temperature properties of hot-rolled structural steels for use in fire engineering studies", *Fire Safety Journal* **13**(1): 27–37.

Kirby, B., Wainman, D., Tomlinson, L. and Peacock, B. (1993) "Natural fires in large scale compartments", British Steel Technical – Fire Research Station Collaborative Project SL/HED/RSC/S11805/1/93/X, British Steel Technical, Swinden Laboratories.

Kirby, B. R. (1997) "Large scale fire tests: the British Steel European collaborative research programme on the BRE 8-storey frame", *Proceedings of the Fifth International Symposium on Fire Safety Science*, Melbourne, Australia, pp. 1129–1140.

Kirby, B. R. and Tomlinson, L. N. (2000) *The Temperature Attained by Unprotected Steelwork in Building Fires*, CORUS Research, Development & Technology.

Kodur, V. K. R. and Wang, Y. C. (2001) "Performance of high strength concrete filled steel columns at ambient and elevated temperatures", in R. Puthli and S. Herion (eds), *Tubular Structures IX*, Lisse: Balkema Publishers.

Kodur, V. K. R. (1998) "Performance of high strength concrete-filled steel columns exposed to fire", *Canadian Journal of Civil Engineering* 25: 975–981.

Kodur, V. K. R. (1999) "Performance based fire resistance design of concrete-filled steel columns", *Journal of Constructional Steel Research* 5: 21–36.

Koike, S., Ooyanagi, N. and Nakamura, K. (1982) "Experimental study on thermal stress within structural steelwork", *Fire Science and Technology* 2(2): 137–150.

Kordina, K. (1989) "Behaviour of composite columns and girders in fire", *Proceedings of the Second International Symposium on Fire Safety Science*, Edinburgh.

Kruppa, J. (1981–82) "Some results on the fire behaviour of external columns", *Fire Safety Journal* 4: 247–257.

Latham, D., Kirby, B., Thomson, G. and Wainman, D. (1986) "The temperatures attained in unprotected structural steelwork in natural fires", Report RSC/7281/10/86, British Steel Technical, Swinden Laboratories.

Law, M. (1970) "A relationship between fire grading and building design and contents", Fire Research Note 901, Joint Fire Research Organization, Building Research Establishment.

Law, M. and O'Brien, T. (1989) *Fire Safety of Bare External Structural Steel*, The Steel Construction Institute, Ascot.

Lawson, R. M. (1990a) "Enhancement of fire resistance of beams by beam to column connections", Technical Report 086, The Steel Construction Institute.

Lawson, R. M. (1990b) "Behaviour of steel beam-to-column connections in fire", *The Structural Engineer* 68(14): 263–271.

Lawson, R. M. (1993) "Building design using cold formed steel sections: fire protection", *SCI Publication* P129, The Steel Construction Institute.

Lawson, R. M. and Newman, G. N. (1990) *Fire Resistant Design of Steel Structures*, The Steel Construction Institute, Ascot.

Lawson, R. M. and Newman, G. M. (1996) "Structural fire design to EC3 & EC4, and comparison with BS 5950", Technical Report, *SCI Publication* 159, The Steel Construction Institute.

Lennon, T. and Simms, C. (1993) "Elevated temperature column tests: results from phase 1", *BRE Note* N201/93, Building Research Establishment.

Lennon, T. (1994) "Elevated Tempertaure column tests: results from phase 2", *BRE Note* N28/94, Building Research Establishment.

Leston-Jones, L. C. (1997) "The influence of semi-rigid connections on the performance of steel framed structures in fire", Ph.D. thesis, University of Sheffield.

Leston-Jones, L. C., Burgess, I. W., Lennon, T. and Plank, R. J. (1997) "Elevated-temperature moment-rotation tests on steelwork connections", *Proceedings of the Institution of Civil Engineers, Structures & Buildings* **122**: 410–419.

Li, G. Q. and Jiang, S. C. (1999) "Prediction to nonlinear behaviour of steel frames subjected to fire", *Fire Safety Journal* **32**: 347–368.

Li, G. Q., Jiang, S. C. and Lin, G. Q. (1999) *Steel Structures: Fire Resistance Calculations and Design*, Beijing: Chinese Construction Industry Publisher.

Lie, T. T. (1974) "Characteristic temperature curves for various fire severities", *Fire Technology* **10**: 315–326.

Lie, T. T. (1980) "New facility to determine fire resistance of columns", *Canadian Journal of Civil Engineering* **7**(3): 551–558.

Lie, T. T. (1992) *Structural Fire Protection*, ASCE Manuals and Reports on Engineering, No. 78, American Society of Civil Engineers.

Lie, T. T. and Chabot, M. (1992) "Experimental studies on the fire resistance of hollow steel columns filled with plain concrete", *Internal Report* No. 611, National Research Council of Canada.

Lie, T. T. and Kodur, V. J. R. (1996) "Fire resistance of steel columns filled with bar-reinforced concrete", *ASCE Journal of Structural Engineering* **122**(1): 30–36.

Liew, J. Y. R., Tan, L. K., Holmaas, T. and Choo, Y. S. (1998) "Advanced analysis for the assessment of steel frames in fire", *Journal of Constructional Steel Research* **47**: 19–45.

Liu, T. C. H. (1988) "Theoretical modeling of steel portal frame behaviour", Ph.D. thesis, Department of Civil Engineering, University of Manchester.

Liu, T. C. H. (1994) "Theoretical modeling of steel bolted connection under fire exposure", *Proceedings of International Conference on Computational Methods in Structural and Geotechnical Engineering Mechanics*, Hong Kong.

Liu, T. C. H. (1996) "Finite element modeling of behaviour of steel beams and connections in fire", *Journal of Constructional Steel Research* **36**(2): 181–199.

Liu, T. C. H. (1998) "Effect of connection flexibility on fire resistance of steel beams", *Journal of Constructional Steel Research* **45**(1): 99–118.

Liu, T. C. H. (1999a) "Fire resistance of unprotected steel beams with moment connections", *Journal of Constructional Steel Research* **51**: 61–77.

Liu, T. C. H. (1999b) "Moment–rotation–temperature characteristics of steel/composite connections", *ASCE Journal of Structural Engineering* **125**(10): 1188–1197.

Ma, Z. C. and Makelainen, P. (1999a) "Behaviour of composite slim floor structures in fire", *ASCE Journal of Structural Engineering* **126**(7): 830–837.

Ma, Z. C. and Makelainen, P. (1999b) "Numerical analysis of steel concrete composite slim floor structures in fire", *Laboratory of Steel Structures Publications* **11**, Helsinki University of Technology.

Ma, Z. C. and Makelainen, P. (2000) "Parametric temperature–time curves of medium compartment fires for structural design", *Fire Safety Journal* **34**: 361–375.

Malhotra, H. L. (1982) *Design of Fire-Resistant Structure*, Surrey University Press.

Martin, D. M. and Moore, D. B. (1997) "Introduction and background to the research programme and major fire tests at BRE Cardington", *Proceedings of the National Steel Construction Conference*, London, May.

Milke, J. A. (1992) "Temperature analysis of structures exposed to fire", *Fire Technology*, May, pp. 184–189.

Mitchell, D. and Cook, W. D. (1983) "Preventing progressive collapse of slab structures", *ASCE Journal of Structural Engineering* **110**(7): 1513–1532.

Myllymaki, J. and Kokkala, M. (2000a) "Fire tests of a steel beam below ceiling exposed to a localized fire", *Proceedings of the Fifth Finnish Steel Structures*, Helsinki University of Technology, Finland, January.

Myllymaki, J. and Kokkala, M. (2000b) "Thermal exposure to a high welded I-beam above a pool fire", *Proceedings of the First International Workshop on Structures in Fire*, Copenhagen, Denmark, June.

Najjar, S. R. (1994) "Three-dimensional analysis of steel frames and subframes in fire", Ph.D. thesis, Department of Civil and Structural Engineering, University of Sheffield, UK.

Najjar, S. R. and Burgess, I. W. (1996) "A non-linear analysis for 3-dimensional steel frames in fire conditions", *Engineering Structures* **18**: 77–89.

Newman, G. M. (1990) *Fire and Steel Construction: the Behaviour of Steel Portal Frames in Boundary Conditions*, 2nd edn, The Steel Construction Institute, Ascot.

Newman, G. M. and Lawson, R. M. (1991) "Fire Resistance of Composite Beams", Technical Report, Publication 109, The Steel Construction Institute.

Newman, G. M. (1995) "Fire resistance of slim floor beams", *Journal of Constructional Steel Research* **33**: 87–100.

Newman, G. M., Robinson, J. T. and Bailey, C. G. (2000) "Fire safety design: a new approach to multi-storey steel-framed buildings", *SCI Publication* P288, The Steel Construction Institute.

Nethercot, D. A. (1985) "Steel beam to column connections – a review of test data", *CIRIA Project Record* **338**.

O'Callaghan, D. J. and O'Connor, M. A. (2000) "Comparison of finite element models of composite steel framed buildings behaviour in fire", *Proceedings of the First International Workshop on Structures in Fire*, Copenhagen, Denmark, June.

Ockleston, A. J. (1955) "Load tests on a three storey reinforced concrete building in Johannesburg", *The Structural Engineer* **33**(10): 304–322.

Ockleston, A. J. (1958) "Arching action in reinforced concrete slabs", *The Structural Engineer* **36**(6): 197–201.

O'Connor, M. A. and Martin, D. M. (1998) "Behaviour of a multi-storey steel framed building subjected to fire attack", *Journal of Constructional Steel Research* **46**(1–3): 295.

Ooyanagi, N., Hirota, M., Nakamura, K. and Kawagoe, K. (1983) "Experimental study on thermal stress within steel frames", *Fire Science and Technonogy* **3**(1): 45–55.

Outinen, J., Kaitila, O. and Makelainen, P. (2000) "A study for development of the design of steel structures in fire conditions", *Proceedings of the 1st International Workshop on Structures in Fire*, Copenhagan, Denmark, pp. 439–444.

Outinen, J. and Makelainen, P. (2001) "Effect of high temperature on mechanical properties of cold-formed structural steel", in R. Puthli and S. Herion (eds), *Tubular Structures IX*, Lisse: Balkema Publishers.

Ove Arup & Partners (OAP) (1985) "Large panel systems – prediction of behaviour in fire tests", A report submitted by OAP to the Building Research Establishment.

Park, R. (1964a) "Ultimate strength of rectangular concrete slabs under short-term uniform loading with edges restrained against lateral movement", *Proceedings of the Institution of Civil Engineers* **28**: 125–150.

Park, R. (1964b) "The ultimate strength and long-term behaviour of uniformly loaded, two-way concrete slabs with partial lateral restraints at all edges", *Magazine of Concrete Research* **16**(48): 139–152.

Park, R. (1964c) "Tensile membrane behaviour of uniformly loaded rectangular reinforced concrete slabs with fully restrained edges", *Magazine of Concrete Research* **16**(46): 39–44.

Park, R. (1965) "The lateral stiffness and strength required to ensure membrane action at the ultimate load of a reinforced concrete slab-and-team floor", *Magazine of Concrete Research* **17**(50): 29–38.

Pchelintsev, A., Hasemi, Y., Wakamatsu, T. and Yokobayashi, Y. (1997) "Experimental and numerical study on the behaviour of a steel beam under ceiling exposed to a localised fire", *Proceedings of the Fifth International Symposium on Fire Safety Science*, Melbourne, March.

Pettersson, O., Magnusson, S. E. and Thor, J. (1976) *Fire Engineering Design of Steel Structures*, Publication 50, Swedish Institute of Steel Construction.

Phan, L. T. and Carino, N. J. (1998) "Review of mechanical properties of HSC at elevated temperatures", *ASCE Journal of Materials in Civil Engineering* **10**(1): 58–64.

Phan, L. T. and Carino, N. J. (2000) "Fire performance of high strength concrete: research needs", *Proceedings of ASCE Congress 2000*, Philadelphia, USA, May.

Piloto, P. A. G. and Vila Real, P. M. M. (2000) "Lateral torsional buckling of steel I-beams in case of fire – experimental evaluation", *Proceedings of the First International Workshop on Structures in Fire*, Copenhagen, Denmark, pp. 95–105.

Plank, R., Burgess, I. and Bailey, C. (1996) "Modelling the behaviour of steel-framed building structures by computer", *Proceedings of the Second Cardington Conference*, Bedfordshire, England.

Plen, E. (1975) "Theoretical and experimental investigation of point set structures", *Document D9*, Swedish Council for Building Research.

Powell-Smith, V. and Billington, M. J. (1999) *The Building Regulations, Explained and Illustrated*, 11th edn, Oxford: Blackwell Science.

Proe, D. J. and Bennetts, I. D. (1994) *Real fire tests in 380 Collins street office enclosure*, BHPR/PPA/R/94/SG021A, BHP Research, Melbourne, Australia.

Purkiss, J. A. (1996) *Fire Safety Engineering Design of Structures*, London: Butterworth-Heinemann.

Quintiere, J. G. (1989a) "Fundamentals of enclosure fire zone models", *Journal of Fire Protection Engineers* **1**(3): 99–119.

Quintiere, J. G. (1989b) "Scaling applications in fire research", *Fire Safety Journal* **15**: 3–29.

Ramachandran, G. (1998) *The Economics of Fire Protection*, London: E & FN Spon.

Ranby, A. (1998) "Structural fire design of thin walled steel sections", *Journal of Constructional Steel Research* **46**(1–3): paper No. 176.

Rhodes, J. (1991) *Design of Cold Formed Steel Structures*, Oxford: Elsevier Applied Science.

Robinson, J. T. and Latham, D. J. (1986) "Fire resistant steel design – the future challenge", in R. D. Anchor, H. J. Malhotra and J. A. Purkiss (eds), *Design of Structures Against Fire*, pp. 225–236.

Rose, P. (1999) "Simulation of steel/concrete composite structures in fire", Ph.D. thesis, University of Sheffield.

Rubert, A. and Schaumann, P. (1986) "Structural steel and plane frame assemblies under fire action", *Fire Safety Journal* 10: 173–184.

Rudolph, K., Richter, E., Hass, R. and Quast, U. (1986) "Principles for calculation of load-bearing and deformation behaviour of composite structural elements under fire action (STABA-F)", *Proceedings of the First International Symposium on Fire Safety Science*, Gaithburg, USA, pp. 301–310.

Ryder, N. L., Wolin, S. D. and Milke, J. A. (2001) "Reduction in fire resistance of columns caused by loss of spray applied fire protection", *Journal of Fire Protection Engineers*.

Saab, H. A. (1990) "Non-linear finite element analysis of steel frames in fire conditions", Ph.D. thesis, University of Sheffield.

Saab, H. A. and Nethercot, D. A. (1991) "Modelling steel frame behaviour under fire conditions", *Engineering Structures* 13: 371–382.

Sakumoto, Y., Okada, T., Yoshida, M. and Tasaka, S. (1993) "Fire resistance of concrete-filled, fire-resistant steel-tube columns", *ASCE Journal of Materials in Civil Engineering* 6(2): 169–184.

Sakumoto, Y., Keira, K., Furumura, F. and Ave, T. (1994) "Tests of fire-resistant bolts and joints", *ASCE Journal of Structural Engineering* 119(11): 3131–3150.

Sakumoto, Y. (1995) "Fire-safe design of modern steel buildings in Japan", *Journal of Constructional Steel Research* 33(1–2): 101–123.

Sakumoto, Y. (1998) "Research on new fire-protection materials and fire-safe design", *ASCE Journal of Structural Engineering* 125(12): 1415–1422.

Sanad, A. M., Rotter, J. M., Usmani, A. S. and O'Connor, M. A. (2000) "Composite beams in large buildings under fire – numerical modelling and structural behaviour", *Fire Safety Journal* 35: 165–188.

Sanad, A. M., Lamont, S., Usmani, A. S. and Rotter, J. M. (2000a) "Structural behaviour in fire compartment under different heating regimes – Part 1: Slab thermal gradients", *Fire Safety Journal* 35: 99–116.

Sanad, A. M., Lamont, S., Usmani, A. S. and Rotter, J. M. (2000b) "Structural behaviour in fire compartment under different heating regimes – Part 2: Slab mean temperatures", *Fire Safety Journal* 35: 117–130.

Sawczuk, A. and Winnicki, L. (1965) "Plastic behaviour of simply supported reinforced concrete plates at moderately large deflections", *International Journal of Solids and Structures* 1: 97–111.

Schleich, J. B., Dotreppe, J. C. and Franssen, J. M. (1986) "Numerical simulation of fire resistance tests on steel and composite structural elements or frames", *Proceedings of the First International Symposium on Fire Safety Science*, NIST, pp. 311–322.

Schleich, J. B. (1996) "A natural fire safety concept for buildings – 1", *Proceedings of the Second Cardington Conference*, Bedford, March.

Sha, W. (1998) "Fire resistance of floors constructed with fire-resistant steels", *ASCE Journal of Structural Engineering* 124(6): 664–670.

Sha, W. and Kelly, F. S. (1999) "A comparison of the mechanical properties of fire-resistant and S275 structural steels", *Journal of Constructional Steel Research* 50: 223–233.

Shepherd, P. (1999) "The performance in fire of restrained columns in steel-framed construction", Ph.D. thesis, University of Sheffield.

Simms, W. I., O'Connor, D. J., Ali, F. and Randall, M. (1995–96) "An experimental investigation on the structural performance of steel columns subjected to elevated temperatures", *Journal of Applied Fire Science* 5(4): 269–284.

Society of Fire Protection Engineers (1995) *SFPE Handbook of Fire Protection Engineering*, 2nd edn, Massachusetts, National Fire Protection Association.

Song, L., Izzuddin, B. A. and Elnashai, A. S. (1995) "Nonlinear analysis of steel frames subjected to explosion and fire loading", *Proceedings of the International Conference on Structural Dynamics, Vibration, Noise and Control*, Hong Kong, December.

Song, L. (1998) "Integrated analysis of steel buildings under fire and explosion", Ph.D. thesis, Department of Civil Engineering, Imperial College, London.

Song, L., Izzuddin, B. A., Elnashai, A. S. and Dowling, P. J. (2000) "An integrated adaptive environment for fire and explosion analysis of steel frames – Part 1: analytical models", *Journal of Constructional Steel Research* 53: 63–85.

Spyrou, S. and Davison, J. B. (2001) "Displacement measurement in studies of steel T-stub connections", *Journal of Constructional Steel Research* 57: 647–659.

Steel Construction Industry Forum (1991) *Investigation of Broadgate phase 8 fire*, The Steel Construction Institute.

Sultan, M. A. (1996) "A model for predicting heat transfer through noninsulated unloaded steel-stud gypsum board wall assemblies exposed to fire", *Fire Technology* 32(3): 239–259.

Talamona, D., Franssen, J. M., Schleich, J. B. and Kruppa, J. (1997) "Stability of steel columns in case of fire: numerical modelling, American Society of Civil Engineers", *Journal of Structural Engineering* 123(6): 713–720.

Talamona, D. and Franssen, J. M. (2000) "New quadrangular shell element in SAFIR", *Proceedings of the First International Workshop on Structures in Fire*, Copenhagen, Denmark.

Talamona, D., Kruppa, J., Franssen, J. M. and Recho, N. (1996) "Factors influencing the behaviour of steel columns exposed to fire", *Journal of Fire Protection Engineers* 8(1): 31–43.

Tan, L. K. (2000) "Advanced analysis for the assessment of steel structures in fire", Ph.D. thesis, Department of Civil Engineering, National University of Singapore.

Taylor, C. (2001) "BS 5950 Part 1: 2000 technical details", *New Steel Construction*, March/April, 25–27.

Taylor, R. (1965) "A note on a possible basis for a new method of ultimate load design of reinforced concrete slabs", *Magazine of Concrete Research* 17(53): 183–186.

Terro, M. J. (1991) "Numerical modelling of thermal and structural response of reinforced concrete structures in fire", Ph.D. thesis, Imperial College of Science and Technology.

Terro, M. J., Khoury, G. and Sullivan, P. (1991) "Critical review of fire-dedicated thermal & structural computer programs", *Fire report submitted to the Building Research Establishment*, Fire Safety Design Consultants, UK.

Thomas, I. R., Bennetts, I. D., Dayawansa, P., Proe, D. J. and Lewis, R. R. (1992) *Fire Tests of the 140 William Street Office Building*, BHPR/ENG/R/92/043/SG2C, BHP Research, Melbourne, Australia.

Thomas, L. C. (1980) *Fundamentals of Heat Transfer*, Englewood Cliffs: Prentice-Hall.

Thomas, P. H., Hinkley, P. L., Theobald, C. R. and Simms, D. L. (1963) "Investigations into the flow of hot gases in roof venting", *Fire Research Technical Paper* No. 7, Joint Fire Research Organization, Building Research Establishment.

Thomas, P. H. and Heselden, A. J. (1972) "Fully developed fires in single compartments, A cooperative programme of the Conseil International Du Batiment (CIB), CIB Report No. 20", *Fire Research Station Note* No. 923, Joint Fire Research Organization, Building Research Establishment.

Thomas, P. H. and Law, M. (1974) "The projection of flames from buildings in fire", *Fire Prevention Science and Technology* 10: 19–26.

Thomas, P. H. (1992) "Fire modeling: a mature technology"? *Fire Safety Journal* 19: 125–140.

Thomas, P. H. and Smith, C. I. (1993) A short guide to the theoretical background to equivalent fire resistance and fire severity, Personal Communication.

Tomecek, D. V. and Milke, J. A. (1993) "A study of the effect of partial loss of protection on the fire resistance of steel columns", *Fire Technology* 29(1): 3–21.

Towler, K., Khoury, G. and Sullivan, P. (1989) "Computer modeling of the effect of fire on large panel structures", Final Report submitted to the Department of Environment by Fire Safety Design Consultants, UK.

Usmani, A. S., Rotter, J. M., Lamont, S., Sanad, A. M. and Gillie, M. (2001) "Fundamental principles of structural behaviour under thermal effects", *Fire Safety Journal* 36: 721–744.

Uy, B. and Bradford, M. A. (1995) "Local buckling of cold formed steel in composite structural elements at elevated temperatures", *Journal of Constructional Steel Research* 34: 53–73.

Van Foeken, R. J. and Snijder, H. H. (1985) "Steel column and frame stability analysis using finite element techniques", *Heron* 30(4).

Vila Real, P. M. M. and Franssen, J. M. (2000) "Lateral torsional buckling of steel I-beams in case of fire – numerical modeling", *Proceedings of the First International Workshop on Structures in Fire*, Copenhagen, Denmark, pp. 71–93.

Wainman, D. E. and Kirby, B. R. (1987) *Compendium of UK Standard Fire Test Data, Unprotected Structural Steel – 1*, Ref. No. RS/RSC/S10328/1/98/B, British Steel Corporation (now Corus), Swinden Laboratories, Rotherham.

Wainman, D. E. and Kirby, B. R. (1988) *Compendium of UK Standard Fire Test Data, Unprotected Structural Steel – 2*, Ref. No. RS/RSC/S1199/8/88/B, British Steel Corporation (now Corus), Swinden Laboratories, Rotherham.

Wang, Y. C. (1992) "A computer program for structural analysis at elevated temperatures", Client Report, Building Research Establishment.

Wang, Y. C. and Moore, D. B. (1992) "Fire resistance of steel frames", in J. L. Clarke, F. K. Garas and G. S. T. Armer (eds), *Structural Design for Hazardous Loads, The Role of Physical Testing*, London: E & FN SPON.

Wang, Y. C. (1993) "Finite element analysis of shell structures in fire", Client Report CR3/93, Building Research Establishment.

Wang, Y. C. (1994a) "Effect of thermal restraint on column behaviour in a frame", *Proceedings of the 4th International Symposium on Fire Safety Science*, Ottawa, Canada, pp. 1055–1066.

Wang, Y. C. (1994b) "Predicting the behaviour of reinforced concrete slabs at elevated temperatures using finite shell element analysis", *Proceedings of the Fourth International Symposium on Fire Safety Science*, Ottawa, Canada, pp. 1077–1088.

Wang, Y. C. and Moore, D. B. (1994) "The effect of frame continuity on the critical temperature of steel columns", *Proceedings of the third international KERENSKY conference on global trends in Structural Engineering*, Singapore, July.

Wang, Y. C. and Moore, D. B. (1995) "Steel frames in fire: analysis", *Engineering Structures* **17**(6): 462–472.

Wang, Y. C., Lennon, T. and Moore, D. B. (1995) "The behaviour of steel frames subject to fire", *Journal of Constructional Steel Research* **35**: 291–322.

Wang, Y., Cooke, G. and Moore, D. (1996) "Large compartment fire tests at Cardington and the Assessment of Eurocode 1", *IABSE Colloquium: Basis of Design and Actions on Structures, Background and Application of Eurocode 1*, Delft, Holland.

Wang, Y. C. (1996) "Tensile membrane action in slabs and its application to the Cardington fire tests", *Proceedings of the 2nd Cardington Conference*, Bedford, UK.

Wang, Y. C. (1997a) "The effects of frame continuity on the behaviour of steel columns under fire conditions and fire resistant design proposals", *Journal of Constructional Steel Research* **41**(1): 93–111.

Wang, Y. C. (1997b) "Effects of structural continuity on fire resistant design of steel columns in non-sway multi-storey frames", *Fire Safety Journal* **28**: 101–106.

Wang, Y. C. (1997c) "Some considerations in the design of unprotected concrete-filled steel tubular columns under fire conditions", *Journal of Constructional Steel Research* **44**(3): 203–223.

Wang, Y. C. (1997d) "Tensile membrane action and the fire resistance of steel framed buildings", *Proceedings of the 5th International Symposium on Fire Safety Science*, Melbourne, Australia, pp. 1117–1128.

Wang, Y. C. (1998a) "Composite beams with partial fire protection", *Fire Safety Journal* **30**: 315–332.

Wang, Y. C. (1998b) "Full scale testing of multi-storey buildings in the large building test facility, Cardington", in D. Dubina, I. Vayas and I. Ungureanu (eds), *New Technologies and Structures in Civil Engineering: Case Studies on Remarkable Constructions*, Editura Orizonturi Universitare, Timisoara, pp. 219–235.

Wang, Y. C. (1999) "The effects of structural continuity on the fire resistance of concrete filled columns in non-sway columns", *Journal of Constructional Steel Research* **50**: 177–197.

Wang, Y. C. and Kodur, V. K. R. (1999) "An approach for calculating the failure loads of unprotected concrete filled steel columns exposed to fire", *Structural Engineering and Mechanics* **7**(2): 127–145.

Wang, Y. C. (2000a) "A simple method for calculating the fire resistance of concrete-filled CHS columns", *Journal of Constructional Steel Research* **54**: 365–386.

Wang, Y. C. (2000b) "An analysis of the global structural behaviour of the Cardington steel-framed building during the two BRE fire tests", *Engineering Structures* **22**: 401–412.

Wang, Y. C. and Kodur, V. K. R. (2000) "Research towards use of unprotected steel structures", *ASCE Journal of Structural Engineering* **126**(12): 1442–1450.

Wang, Y. C. and Davies, J. M. (2000) "Design of thin-walled steel channel columns in fire using Eurocode 3 Part 1.3", *Proceedings of the First International Workshop on Structures in Fire*, Copenhagen, Denmark, pp. 181–193.

Wang, Y. C. (2001a) "Temperatures in protected steel in natural fires", *Fire Safety Journal* (submitted for publication).

Wang, Y. C. (2001b) "Post-buckling of axially restrained steel columns under axial loads in fire", *Journal of Constructional Steel Research* (submitted for publication).

Wang, Y. C. (2001c) "Bending moment transfer in restrained concrete filled columns in fire" (unpublished results).

Wickstrom, U. (1979) "TASEF-2, A computer program for temperature analysis of structures exposed to fire", Lund Institute of Technology.

Wickstrom, U. (1982) "Temperature calculation of insulated steel columns exposed to natural fire", *Fire Safety Journal* 4(4): 219–225.

Wickstrom, U. (1985) "Temperature analysis of heavily-insulated steel structures exposed to fire", *Fire Safety Journal* 9: 281–285.

Wickstrom, U. (1988) "A proposal regarding temperature measurements in fire test furnaces", *Fire Technology Technical Report* SP-RAPP 1986:17, Swedish National Testing Institute.

Wickstrom, U. (1989) "The plate thermometer – a simple instrument for reaching harmonized fire resistance tests", *Fire Technology SP Report* 1989:03, Swedish National Testing Institute.

Wickstrom, U. and Hermodsson, T. (1997) "Comments on paper by Kay, Kirby and Preston 'calculation of the heating rate of an unprotected steel member in a standard fire resistance test'", *Fire Safety Journal* 29: 337–343.

Willam, J. J. (1974) "Constitutive model for triaxial behaviour of concrete", *Proceedings of IABSE Seminar*, Bergamo, Italy.

Witteveen, J. and Twilt, L. (1981–82) "A critical view of the results of standard fire resistance tests on steel columns", *Fire Safety Journal* 4: 259–270.

Wong, M. B. (2001) "Elastic and plastic methods for numerical modelling of steel structures subject to fire", *Journal of Constructional Steel Research* 57: 1–14.

Wood, R. H. (1961) *Plastic and Elastic Design of Slabs and Plates*, London: Thames and Hudson.

Wood, R. H. (1974) "A new approach to column design, with special reference to restrained steel stanchions", *Building Research Establishment Report*, Building Research Establishment.

Yan, Z. H. and Holmstedt, G. (1997) "CFD and experimental studies of room fire growth on wall lining materials", *Fire Safety Journal* 27: 201–238.

Yandzio, E., Dowling, J. J. and Newman, G. N. (1996) "Structural fire design: off-site applied thin film intumescent coatings", *SCI Publication* 160, The Steel Construction Institute.

Yu, W. W. (1991), *Cold-Formed Steel Design*, 2nd edn, New York: Wiley-Interscience.

Zhao, B. and Kruppa, J. (1995) *Fire resistance of composite slabs with profiled steel sheet and of composite steel concrete beams, Part 2: Composite beams*, Final report, ECSC – agreement No. 7210 SA 509, CTICM, France.

Zienkiewicz, O. C. and Taylor, R. L. (1991) *The Finite Element Method*, 4th edn, New York: McGraw-Hill.

Author index

Subject index

Milton Keynes UK
Ingram Content Group UK Ltd.
UKHW021630071024
449327UK00020BA/1265